Library Recommendations for Undergraduate Mathematics

Lynn Arthur Steen, Editor

THE MATHEMATICAL ASSOCIATION OF AMERICA

MAA Notes and Reports Series

The MAA Notes and Reports Series, started in 1982, addresses a broad range of topics and themes of interest to all who are involved with undergraduate mathematics. The volumes in this series are readable, informative, and useful, and help the mathematical community keep up with developments of importance to mathematics.

MAA Notes

1. Problem Solving in the Mathematics Curriculum,
   Committee on the Teaching of Undergraduate Mathematics,
   a subcommittee of the Committee on the Undergraduate Program in Mathematics, *Alan H. Schoenfeld,* Editor.
2. Recommendations on the Mathematical Preparation of Teachers,
   Committee on the Undergraduate Program in Mathematics, Panel on Teacher Training.
3. Undergraduate Mathematics Education in the People's Republic of China,
   Lynn A. Steen, Editor.
4. Notes on Primality Testing and Factoring,
   Carl Pomerance.
5. American Perspectives on the Fifth International Congress on Mathematical Education,
   Warren Page, Editor.
6. Toward a Lean and Lively Calculus,
   Ronald G. Douglas, Editor.
7. Undergraduate Programs in the Mathematical and Computer Sciences: 1985–86,
   D. J. Albers, R. D. Anderson, D. O. Loftsgaarden, Editors.
8. Calculus for a New Century,
   Lynn A. Steen, Editor.
9. Computers and Mathematics: The Use of Computers in Undergraduate Instruction,
   Committee on Computers in Mathematics Education, D. A. Smith, G. J. Porter, L. C. Leinbach, and R. H. Wenger, Editors.
10. Guidelines for the Continuing Mathematical Education of Teachers,
    Committee on the Mathematical Education of Teachers.
11. Keys to Improved Instruction by Teaching Assistants and Part-Time Instructors,
    Committee on Teaching Assistants and Part-Time Instructors, Bettye Anne Case, Editor.
12. The Use of Calculators in the Standardized Testing of Mathematics,
    John Kenelly, Editor, published jointly with The College Board.
13. Reshaping College Mathematics,
    Committee on the Undergraduate Program in Mathematics, Lynn A. Steen, Editor.
14. Mathematical Writing,
    by *Donald E. Knuth, Tracy Larrabee, and Paul M. Roberts.*
15. Discrete Mathematics in the First Two Years,
    Anthony Ralston, Editor.
16. Using Writing to Teach Mathematics,
    Andrew Sterrett, Editor.
17. Priming the Calculus Pump: Innovations and Resources,
    Committee on Calculus Reform and the First Two Years,
    a subcommittee of the Committee on the Undergraduate Program in Mathematics, *Thomas W. Tucker,* Editor.

18. Models for Undergraduate Research in Mathematics,
Lester Senechal, Editor.

19. Visualization in Teaching and Learning Mathematics,
Committee on Computers in Mathematics Education, Steve Cunningham and Walter S. Zimmermann, Editors.

20. The Laboratory Approach to Teaching Calculus,
L. Carl Leinbach et al., Editors.

21. Perspectives on Contemporary Statistics,
David C. Hoaglin and David S. Moore, Editors.

MAA Reports

A Curriculum in Flux: Mathematics at Two-Year Colleges,
Subcommittee on Mathematics Curriculum at Two-Year Colleges, a joint committee of the MAA and the American Mathematical Association of Two-Year Colleges, *Ronald M. Davis,* Editor.

A Source Book for College Mathematics Teaching,
Committee on the Teaching of Undergraduate Mathematics, Alan H. Schoenfeld, Editor.

A Call for Change:
Recommendations for the Mathematical Preparation of Teachers of Mathematics,
Committee on the Mathematical Education of Teachers, James R. C. Leitzel, Editor.

Library Recommendations for Undergraduate Mathematics,
CUPM ad hoc Subcommittee, Lynn A. Steen, Editor.

First Printing

ISBN 0-88385-076-1
Library of Congress Catalog
Card Number 91-067940
Printed in the United States of America

Table of Contents

Preface

Strong libraries are of vital importance to an effective undergraduate environment in which students are expected to acquire the disposition for life-long learning. To accomplish this goal, colleges must develop curricula—in mathematics as in any other subject—that encourage students to learn to use a library. However, before that can happen, colleges must develop libraries worth using.

Unfortunately, undergraduate students rarely use library resources in their mathematical studies. Conventional curricula and common teaching practices rely almost exclusively on textbooks—despite substantial evidence that more engaging assignments based on richer resources lead to better understanding. Moreover, typical undergraduate libraries have very thin and eccentric mathematics collections, partly because many librarians have little knowledge of mathematics and many mathematicians have little interest in libraries. Collections are typically shaped more by the happenstance of research interests of new faculty than by careful planning that emphasizes balance of the collection as a whole. This process yields spotty collections that are often inappropriate for undergraduate education.

Building a strong undergraduate library is much like building a strong undergraduate curriculum. One must plan for the whole while ensuring balance of all the parts. In building libraries as in planning curricula, one must make tough choices. As not every good course can be taught, neither can every good book be purchased. Course-related library assignments are an essential bridge between textbook-based homework and undergraduate research experiences that are so crucial to career choices made during college years. Moreover, research experiences for undergraduates require strong libraries that reflect the best of current mathematical practice.

The current national imperative for revitalizing mathematics education adds a special sense of urgency to the need for balanced, contemporary undergraduate mathematics collections. A good undergraduate library provides resources for students to broaden their mathematical horizons; opportunities for assignments that move beyond mere exercises to writing, reading, and research; and sources for faculty who are engaged in scholarship and curriculum development. Library improvement in the mathematical sciences is a necessary (but by no means sufficient) step along the path towards curriculum reform.

Background

Twenty-five years ago the Mathematical Association of America addressed a similar need by preparing *Basic Library Lists* for two-year and four-year colleges. The goal then was to encourage all colleges to build mathematics collections adequate to ensure that no student would be denied access to appropriate mathematical materials. The four-year college *List* was revised once, but even that revision is now fifteen years old. Since the *Basic Library List* was last revised, over 15,000 books have been published in the mathematical sciences, many reflecting topics and specialities that have only recently become prominent.

Mathematics itself has undergone several major changes. Computers have emerged as a significant force in mathematics, introducing algorithmic thinking, discrete mathematics, data analysis, and mathematical logic as important and growing parts of the mathematical sciences. The increasing applicability of mathematics has created new demand for courses in mathematical modelling, in statistics, and in mathematical biology. The joining of applied mathematics with computing has created whole new disciplines on the periphery of mathematics, such as the theory of computation, simulation, scientific computation, and dynamical systems.

Powerful new methods of computational and applied mathematics have the potential to attract many good students to careers in mathematics. Unfortunately, these advances are often invisible to students at the early, formative stage of their college studies, since all they see in introductory courses are calculus textbooks. Because of the long delay for modern topics to get into regular texts, it is very important that colleges be encouraged to develop instructional styles that introduce beginning students not only to textbooks, but also to library resources. For that to make educational sense, it is important that libraries use their limited budgets to build collections that will stimulate the interests of all prospective mathematics students, both those who might be attracted to major in a mathematical science as well as the many others who study mathematics for different purposes.

Objectives

Library Recommendations for Undergraduate Mathematics contains approximately 3000 titles arranged in 25 chapters and 230 sections. Books have been selected according to several different objectives:

- To provide students with introductory material in various fields of the mathematical sciences that might not be part of their curriculum;
- To provide reading material that is collateral to regular courses;
- To extend an undergraduate curriculum with introductory post-graduate sources;
- To provide faculty with reference material that is relevant to their teaching and that will help broaden their mathematical expertise;
- To provide resources for independent study and undergraduate experiences in research;
- To provide general readers with clear and lively exposition about the mathematical sciences and their applications;
- To ensure that college libraries have important reference works and classic sources in the mathematical sciences;
- To provide students in disciplines that use the mathematical sciences with appropriate references.

Whereas the former *Basic Library List* included about 1000 volumes, the present volume includes three times as many titles. This growth is both appropriate and necessary, given the dramatic increase in the breadth and applicability of the mathematical sciences. To help libraries on limited budgets, books have been marked with asterisks to indicate priority:

*** *Essential* (approximately 200 titles)
** *Highly recommended* (approximately 400 titles)
* *Recommended* (approximately 800 titles)
Listed (approximately 1600 titles)

The final section contains recommendations for periodical and journal subscriptions in the mathematical sciences, marked in similar priority categories.

By its very nature, *Library Recommendations for Undergraduate Mathematics* offers a comprehensive survey of the best literature in the mathematical sciences. Although intended primarily for libraries, the volume is also an excellent source of inspiration for self-study by anyone wishing to explore a new field or to catch up on the recent literature in a subject. The volume provides a truly panoramic window on the world of mathematics.

Procedures

This volume was prepared under the supervision of a special subcommittee of the MAA's Committee on the Undergraduate Program in Mathematics (CUPM). Primary judgments for titles to be included in each chapter were provided by over 100 college and university mathematics faculty working in 25 teams according to subject area. Initial nominations were drawn primarily from Telegraphic Reviews published in the *American Mathematical Monthly* and from the earlier *Basic Library List*. These initial nominations numbered in excess of 15,000 titles. After preliminary screening, titles were arranged by topic to provide 25 lists totaling approximately 4500 nominations to the discipline review teams. These teams then added another 1500 nominations before beginning the major task of establishing priorities. The recommendations published here represent approximately half of the titles originally considered by the review teams.

The process of review has been long and detailed, requiring not only the melding of conflicting opinions by many different experts but also the checking of countless bibliographic details. The result is an amalgam of expert opinion, not a consensus of our many reviewers. Many judgments reflected in these *Recommendations* are shared by some reviewers but not by others. Our goal was not to make a list of "best books," but to produce a structured set of recommendations that would be useful to undergraduate libraries, that would ensure balance by topic, and that would maintain reasonable limits on the number of recommendations.

Within these constraints, certain informal principles guided the many difficult choices that were required. Routine textbooks were given lower priority than other monographs because it is generally not wise for a library to devote scarce resources to ordinary student texts. In many subjects with dozens of comparable texts, we list a few titles to ensure coverage and omit many others that are essentially similar. In such cases our choices are not primarily a reflection of different quality but of desire to avoid needless duplication. We also tilted choices in favor of recent titles, since this is an area where library committees may be in greatest need of advice. Many old classics which are of great value, especially to libraries that already have them, are not included in our recommendations because new money would probably be more wisely spent on newer books.

Wise use of these *Recommendations* requires judgement and dialogue between librarians and mathematicians. Many good and valuable books do not appear here, some being cut primarily to meet our self-imposed requirements of total length. Different libraries will view differently the relative priorities of textbooks, of references books, or of advanced monographs. In many situations, libraries may already have comparable books that are not on this list but that nevertheless provide important and equivalent coverage of certain areas. While the collection recommended here does constitute an excellent mathematics library, so would collections with other shapes and other titles. Limited library resources should be used not just to replicate this list as a whole, but to build strength in areas where coverage is thin.

Bibliographic Details

Mathematics is a multi-facted discipline with innumerable cross-linkages among widely separated specialties. These connections makes the subject fascinating even as they make the task of the bibliographer frustrating. Many titles could easily be listed in several different sections; some seem to fit no section at all. Notwithstanding the difficulties of logically neat classification, titles are listed only once, in some sub-area in which they seem to fit. No cross references have been used, since once begun such a task could go on almost without end.

Each title is listed with spare but adequate bibliographic detail, typically just author(s), title, publisher, and date. Other details such as translators, book series, or special editions are normally omitted. For books that have been revised or reprinted, we normally give only the most recent title with just two dates (and publishers): the first, and the most recent. We leave to librarians and their mathematics department advisors the subtle question of balancing the purchase of new editions against the purchase of new titles.

No attempt has been made to indicate whether books recommended here are still in print. Many are, and many are not. The thriving publication of reprints renders such information ephemeral, outdated before it would appear. Libraries interested in obtaining out-of-print titles can often secure them through used book markets. Perhaps the very presence of an old title among these *Recommendations* may cause some old classics to be reprinted.

The author index lists each author (or editor) named in the main bibliography together with title of the book and the section in which it appears. This allows anyone checking for particular titles to readily locate them in the main list, where full bibliographic details are provided.

Acknowledgements

Many people helped with this volume, some of whom I know only through the ubiquitous medium of e-mail. Twenty-five subject-area team leaders undertook the primary responsibility of organizing and amalgamating the diverse recommendations pertaining to each chapter of the bibliography. They were assisted by scores of reviewers who provided advice on various parts of the manuscript. Names of those who helped with various stages of the work are listed on the following pages. Everyone who may benefit from this volume owes these hard-working reviewers a real debt of gratitude.

Policy for the volume was set by a Steering Committee which was established by MAA as an *ad hoc* Subcommittee of CUPM. Advice from members of the Steering Committee was valuable at all stages of the process. I especially appreciated their counsel at key points where conflicting recommendations regarding policy needed to be resolved.

Typing and checking of innumerable bibliographic details was carried out with unfailing good humor over a period of more than two years by Mary Kay Peterson, who has also prepared every Telegraphic Review that has appeared in the *Monthly* for the last twenty years. This volume would never have been possible without the consistency provided by her careful attention to details, large and small.

Financial support for preparation of this long-overdue set of recommendations has been provided by grants from the National Science Foundation and the Exxon Education Foundation. Their support made it possible both to prepare the volume and to ensure its distribution to undergraduate libraries.

Despite numerous efforts to check and proof-read this set of recommendations, among the thousands of details there inevitably remain dozens of errors or inconsistencies—a missing edition here, a wrong date there. More substantial errors undoubtedly also occur, improperly categorized titles being the most likely. I hope that these occasional blemishes do not obscure the central purpose and value of the *Recommendations* as a whole: to provide colleges and universities with useful, contemporary guidance concerning the development of the mathematical sciences collection of an undergraduate library.

Lynn Arthur Steen
St. Olaf College
October 1991

Steering Committee

DONALD J. ALBERS, Associate Director, Mathematical Association of America.

NANCY D. ANDERSON, Mathematics Librarian, University of Illinois, Urbana, IL.

PAUL J. CAMPBELL, Beloit College, Beloit, WI.

CHARLES R. HAMPTON, College of Wooster, Wooster, OH.

LYNN ARTHUR STEEN, St. Olaf College, Northfield, MN.

MARCIA P. SWARD, Executive Director, Mathematical Association of America.

ALAN C. TUCKER, SUNY at Stony Brook, Stony Brook, NY.

ANN E. WATKINS, California State University, Northridge, CA.

Team Leaders

GEORGE E. ANDREWS, Pennsylvania State University, University Park, PA. *(Number Theory)*

RON BARNES, University of Houston, Downtown, Houston, TX. *(Probability)*

JERRY L. BONA, Pennsylvania State University, University Park, PA. *(Physical Sciences)*

BARRY A. CIPRA, Northfield, MN. *(General)*

RAY E. COLLINGS, Tri-County Technical College, Pendleton, SC. *(Vocational Mathematics)*

R. STEPHEN CUNNINGHAM, California State University at Stanislaus, Turlock, CA. *(Computer Science)*

JOHN A. DOSSEY, Illinois State University, Normal, IL. *(Education)*

UNDERWOOD DUDLEY, DePauw University, Greencastle, IN. *(Recreational Mathematics)*

SHELDON P. GORDON, Suffolk County Community College, Selden, NY. *(Two-Year Colleges)*

MEYER JERISON, Purdue University, West Lafayette, IN. *(Analysis)*

BARBARA A. JUR, Macomb Community College, Warren, MI. *(Business Mathematics)*

GENEVIEVE M. KNIGHT, Coppin State College, Baltimore, MD. *(School Mathematics)*

SIMON A. LEVIN, Cornell University, Ithaca, NY. *(Life Sciences)*

R. BRUCE LIND, University of Puget Sound, Tacoma, WA. *(Statistics)*

DANIEL P. MAKI, Indiana University, Bloomington, IN. *(Operations Research)*

JOSEPH MALKEVITCH, York College of CUNY, Jamaica, NY. *(Geometry)*

WALTER E. MIENTKA, University of Nebraska, Lincoln, NE. *(High School Libraries)*

KENNETH C. MILLETT, University of California, Santa Barbara, CA. *(Topology)*

HAL G. MOORE, Brigham Young University, Provo, UT. *(Algebra)*

YVES NIEVERGELT, Eastern Washington University, Cheney, WA. *(Numerical Analysis)*

INGRAM OLKIN, Stanford University, Stanford, CA. *(Social Sciences)*

VERA PLESS, University of Illinois at Chicago, Chicago, IL. *(Discrete Mathematics)*

MARIAN POUR-EL, University of Minnesota, Minneapolis, MN. *(Foundations)*

SAMUEL M. RANKIN III, Worcester Polytechnic Institute, Worcester, MA. *(Differential Equations)*

FREDERICK RICKEY, Bowling Green State University, Bowling Green, OH. *(History)*

JIMMY L. SOLOMON, Mississippi State University, Mississippi State, MS. *(Linear Algebra)*

ALVIN SWIMMER, Arizona State University, Tempe, AZ. *(Calculus)*

Reviewers

JOE ALBREE, Auburn University at Montgomery, Montgomery, AL.

RICHARD ASKEY, University of Wisconsin, Madison, WI.

ED BENDER, J. Sargeant Reynolds Community College, Richmond, VA.

BRUCE C. BERNDT, University of Illinois, Urbana, IL.

WILLIAM D. BLAIR, Northern Illinois University, DeKalb, IL.

WILLIAM E. BOYCE, Rensselaer Polytechnic Institute, Troy, NY.

ANNE E. BROWN, Saint Mary's College, Notre Dame, IN.

GARY BROWN, College of St. Benedict, St. Joseph, MN.

BARRY BRUNSON, Western Kentucky University, Bowling Green, KY.

JOE P. BUHLER, Reed College, Portland, OR.

DONALD BUSHAW, Washington State University, Pullman, WA.

THOMAS J. CARTER, California State University at Stanislaus, Turlock, CA.

J. KEVIN COLLIGAN, National Security Agency, Fort Meade, MD.

MIRIAM P. COONEY, Saint Mary's College, Notre Dame, IN.

ELLIS CUMBERBATCH, Claremont Graduate School, Claremont, CA.

JOHN DAUGHTRY, East Carolina University, Greenville, NC.

NATHANIEL DEAN, Bell Communications Research, Morristown, NJ.

KEITH DEVLIN, Colby College, Waterville, ME.

MARJORIE ENNEKING, Portland State University, Portland, OR.

MARCUS W. FELDMAN, Stanford University, Stanford, CA.

J. CHRIS FISHER, University of Regina, Regina, Saskatchewan, Canada.

BEN A. FUSARO, Salisbury State University, Salisbury, MD.

STEVE GALOVICH, Carleton College, Northfield, MN.

ALAN J. GOLDMAN, Johns Hopkins University, Baltimore, MD.

RONALD J. GOULD, Emory University, Atlanta, GA.

SANDY GRABINER, Pomona College, Claremont, CA.

PAUL R. HALMOS, Santa Clara University, Santa Clara, CA.

JAMES L. HARTMAN, College of Wooster, Wooster, OH.

BRYAN HEARSEY, Lebanon Valley College, Annville, PA.

M. KATHLEEN HEID, Pennsylvania State University, University Park, PA.

ISOM HERRON, Howard University, Washington, DC.

LINDA HILL, Idaho State University, Pocatello, ID.

PAT HIRSCHY, Delaware Technical and Community College, Wilmington, DE.

FREDERICK HOFFMAN, Florida Atlantic University, Boca Raton, FL.

MARY ANN HOVIS, Lima Technical College, Lima, OH.

THOMAS W. HUNGERFORD, Cleveland State University, Cleveland, OH.

ELEANOR GREEN JONES, Norfolk State University, Norfolk, VA.

ROBERT E. KENNEDY, Central Missouri State University, Warrensburg, MO.

MARTHA J. KIRKER, Iowa State University, Ames, IA.

SHANTHA KRISHNAMACHARI, Borough of Manhattan Community College, New York, NY.

DONALD KRUG, Northern Kentucky University, Highland Heights, KY.

CAROL W. KUBLIN, SUNY College of Agriculture and Technology, Cobleskill, NY.

STEPHEN KUHN, University of Tennessee at Chattanooga, Chattanooga, TN.

JAMES M. LANDWEHR, AT&T Bell Labs, Murray Hill, NJ.

PETER A. LINDSTROM, North Lake College, Irving, TX.

ANTHONY LOBELLO, Allegheny College, Meadville, PA.

WILLIAM LUCAS, Claremont Graduate School, Claremont, CA.

MICHAEL MCASEY, Bradley University, Peoria, IL.

DUSA MCDUFF, SUNY at Stony Brook, Stony Brook, NY.

GLENN MILLER, Borough of Manhattan Community College, New York, NY.

ROGERS J. NEWMAN, Southern University, Baton Rouge, LA.

ARNOLD OSTEBEE, St. Olaf College, Northfield, MN.

PETER PAPPAS, Vassar College, Poughkeepsie, NY.

MARY R. PARKER, Austin Community College, Austin, TX.

WALTER M. PATTERSON III, Lander College, Greenwood, SC.

MEE SEE PHUA, University of the District of Columbia, Washington, DC.

JAMES D. REID, Wesleyan University, Middletown, CT.

ROBERT O. ROBSON, Oregon State University, Corvallis, OR.

THOMAS A. ROMBERG, University of Wisconsin, Madison, WI.

FRANCES ROSAMOND, National University, San Diego, CA.

PETER ROSS, Santa Clara University, Santa Clara, CA.

SHARON CUTLER ROSS, DeKalb College, Clarkston, GA.

JAMES L. ROVNYAK, University of Virginia, Charlottesville, VA.

RICHARD L. SCHEAFFER, University of Florida, Gainesville, FL.

JOHN SCHUE, Macalester College, St. Paul, MN.

RAE MICHAEL SHORTT, Wesleyan University, Middletown, CT.

TOM SIBLEY, St. John's University, Collegeville, MN.

MARTHA J. SIEGEL, Towson State University, Towson, MD.

STEPHANIE SLOYAN, Georgian Court College, Lakewood, NJ.

CARL R. SPITZNAGEL, John Carroll University, Cleveland, OH.

ELIZABETH J. TELES, Montgomery College, Takoma Park, MD.

ALLEN TUCKER, Bowdoin College, Brunswick, ME.

THOMAS TUCKER, Colgate University, Hamilton, NY.

WILLIAM C. WATERHOUSE, Penn State University, University Park, PA.

JOAN WYZKOSKI WEISS, Fairfield University, Fairfield, CT.

JULIAN WEISSGLASS, Mathematical Sciences Education Board, Washington, DC.

CAROL WOOD, Wesleyan University, Middletown, CT.

1 General

1.1 General Anthologies

** ALEKSANDROV, A.D.; KOLMOGOROV, ANDREI N.; AND LAVRENT'EV, M.A., EDS. *Mathematics: Its Content, Methods, and Meaning,* 3 Vols. Cambridge, MA: MIT Press, 1969.

BEHNKE, H., *et al.*, EDS. *Fundamentals of Mathematics,* 3 Vols. Cambridge, MA: MIT Press, 1974, 1983.

* BOEHM, GEORGE A.W. *The Mathematical Sciences: A Collection of Essays.* Cambridge, MA: MIT Press, 1969.

CAMPBELL, DOUGLAS M. AND HIGGINS, JOHN C., EDS. *Mathematics: People, Problems, Results,* 3 Vols. Belmont, CA: Wadsworth, 1984.

KAPUR, J.N. *Fascinating World of Mathematical Sciences,* 3 Vols. New Delhi: Mathematical Sciences Trust Society, 1989.

LE LIONNAIS, F., ED. *Great Currents of Mathematical Thought,* 2 Vols. Mineola, NY: Dover, 1971.

*** NEWMAN, JAMES R. *The World of Mathematics,* 4 Vols. New York, NY: Simon and Schuster, 1956; Redmond, WA: Microsoft Press, 1988.

SAATY, THOMAS L. AND WEYL, F. JOACHIM, EDS. *The Spirit and the Uses of the Mathematical Sciences.* New York, NY: McGraw-Hill, 1969.

* SCIENTIFIC AMERICAN. *Mathematics in the Modern World.* New York, NY: W.H. Freeman, 1968.

** STEEN, LYNN ARTHUR, ED. *Mathematics Today: Twelve Informal Essays.* New York, NY: Springer-Verlag, 1978, 1984.

STEEN, LYNN ARTHUR, ED. *Mathematics Tomorrow.* New York, NY: Springer-Verlag, 1981.

1.2 Elementary Exposition

ASIMOV, ISAAC. *Asimov on Numbers.* New York, NY: Doubleday, 1977.

BELL, ERIC T. *Mathematics: Queen and Servant of Science.* Washington, DC: Mathematical Association of America, 1987.

BUNCH, BRYAN H. *Mathematical Fallacies and Paradoxes.* New York, NY: Van Nostrand Reinhold, 1982.

*** COURANT, RICHARD AND ROBBINS, H. *What is Mathematics?* New York, NY: Oxford University Press, 1941.

DANTZIG, TOBIAS. *Number, The Language of Science,* Fourth Edition. New York, NY: Free Press, 1967.

** DAVIS, PHILIP J. AND HERSH, REUBEN. *Descartes' Dream: The World According to Mathematics.* San Diego, CA: Harcourt Brace Jovanovich, 1986; New York, NY: Penguin, 1988.

** DAVIS, PHILIP J. AND HERSH, REUBEN. *The Mathematical Experience.* New York, NY: Birkhäuser, 1980.

** DEVLIN, KEITH J. *Mathematics: The New Golden Age.* New York, NY: Penguin, 1988.

* EKELAND, IVAR. *Mathematics and the Unexpected.* Chicago, IL: University of Chicago Press, 1988.

FLEGG, H. GRAHAM. *Numbers: Their History and Meaning.* New York, NY: Schocken, 1983.

GÅRDING, LARS. *Encounter with Mathematics.* New York, NY: Springer-Verlag, 1977.

*** GLEICK, JAMES. *Chaos: Making a New Science.* New York, NY: Viking Press, 1987.

GUILLEN, MICHAEL. *Bridges to Infinity: The Human Side of Mathematics.* Boston, MA: Houghton Mifflin, 1983.

*** HARDY, G.H. *A Mathematician's Apology,* Revised Edition. New York, NY: Cambridge University Press, 1967, 1969.

* HOFFMAN, PAUL. *Archimedes' Revenge: The Joys and Perils of Mathematics.* New York, NY: W.W. Norton, 1988.

*** HOFSTADTER, DOUGLAS R. *Gödel, Escher, Bach: An Eternal Golden Braid.* New York, NY: Basic Books, 1979.

HOFSTADTER, DOUGLAS R. *Metamagical Themas: Questing for the Essence of Mind and Pattern.* New York, NY: Basic Books, 1985.

HOGBEN, LANCELOT. *Mathematics for the Million.* London: George Allen and Unwin, 1936; England: Merlin Press, 1989.

* HONSBERGER, ROSS. *Ingenuity in Mathematics.* Washington, DC: Mathematical Association of America, 1975.

** HONSBERGER, ROSS. *Mathematical Gems,* 3 Vols. Washington, DC: Mathematical Association of America, 1973–85.

* HONSBERGER, ROSS. *Mathematical Morsels.* Washington, DC: Mathematical Association of America, 1978.

* HONSBERGER, ROSS. *Mathematical Plums.* Washington, DC: Mathematical Association of America, 1979.

* HONSBERGER, ROSS. *More Mathematical Morsels.* Washington, DC: Mathematical Association of America, 1991.

KAHN, DAVID. *Kahn On Codes: Secrets of the New Cryptology.* New York, NY: Macmillan, 1983.

** KAPPRAFF, JAY. *Connections: The Geometric Bridge Between Art and Science.* New York, NY: McGraw-Hill, 1991.

KHURGIN, YA. *Did You Say Mathematics?* Moscow: MIR, 1974.

* KLINE, MORRIS. *Mathematics and the Search for Knowledge.* New York, NY: Oxford University Press, 1985.

* KLINE, MORRIS. *Mathematics: The Loss of Certainty.* New York, NY: Oxford University Press, 1980.

LITTLEWOOD, DUDLEY E. *The Skeleton Key of Mathematics.* New York, NY: Harper and Row, 1960.

** MORRISON, PHILIP AND MORRISON, PHYLIS. *Powers of Ten.* New York, NY: W.H. Freeman, 1982.

*** NIVEN, IVAN M. *Maxima and Minima Without Calculus.* Washington, DC: Mathematical Association of America, 1981.

** PAULOS, JOHN ALLEN. *Beyond Numeracy: Ruminations of a Numbers Man.* New York, NY: Alfred A. Knopf, 1991.

PAULOS, JOHN ALLEN. *Innumeracy: Mathematical Illiteracy and Its Consequences.* New York, NY: Hill and Wang, 1988.

PEDOE, DAN. *The Gentle Art of Mathematics.* New York, NY: Macmillan, 1963.

** PENROSE, ROGER. *The Emperor's New Mind: Concerning Computers, Minds, and The Laws of Physics.* New York, NY: Oxford University Press, 1989.

*** PETERSON, IVARS. *Islands of Truth.* New York, NY: W.H. Freeman, 1991.

*** PETERSON, IVARS. *The Mathematical Tourist: Snapshots of Modern Mathematics.* New York, NY: W.H. Freeman, 1988.

* RADEMACHER, HANS AND TOEPLITZ, OTTO. *The Enjoyment of Mathematics.* Princeton, NJ: Princeton University Press, 1957.

** RADEMACHER, HANS. *Higher Mathematics from an Elementary Point of View.* New York, NY: Birkhäuser, 1983.

* RUCKER, RUDY. *Infinity and the Mind: The Science and Philosophy of the Infinite.* New York, NY: Birkhäuser, 1982.

SCHOENBERG, I.J. *Mathematical Time Exposures.* Washington, DC: Mathematical Association of America, 1982.

* STEINHAUS, HUGO. *Mathematical Snapshots,* Third Edition. New York, NY: Oxford University Press, 1969, 1983.

* STEWART, IAN. *Concepts of Modern Mathematics.* New York, NY: Penguin, 1975.

** STEWART, IAN. *Does God Play Dice? The Mathematics of Chaos.* New York, NY: Penguin, 1989.

** STEWART, IAN. *The Problems of Mathematics.* New York, NY: Oxford University Press, 1987.

WHITEHEAD, ALFRED NORTH. *An Introduction to Mathematics.* New York, NY: Oxford University Press, 1958.

1.3 Advanced Surveys

** ABBOTT, J.C., ED. *The Chauvenet Papers: A Collection of Prize-Winning Expository Papers in Mathematics,* 2 Vols. Washington, DC: Mathematical Association of America, 1978.

* BOCHNER, SALOMON. *The Role of Mathematics in the Rise of Science.* Princeton, NJ: Princeton University Press, 1981.

* BROWDER, FELIX E., ED. *Mathematical Developments Arising from Hilbert Problems.* Providence, RI: American Mathematical Society, 1976.

* DIEUDONNÉ, JEAN AND MACDONALD, I.G. *A Panorama of Pure Mathematics as Seen by N. Bourbaki.* New York, NY: Academic Press, 1982.

DÖRRIE, HEINRICH. *100 Great Problems of Elementary Mathematics.* Mineola, NY: Dover, 1965.

* JÄGER, W.; MOSER, J.; AND REMMERT, REINHOLD, EDS. *Perspectives in Mathematics.* New York, NY: Birkhäuser, 1984.

KAC, MARK; ROTA, GIAN-CARLO; AND SCHWARTZ, JACOB T. *Discrete Thoughts: Essays on Mathematics, Science, and Philosophy.* New York, NY: Birkhäuser, 1986.

KOCH, HELMUT. *Introduction to Classical Mathematics I: From the Quadratic Reciprocity Law to the Uniformization Theorem.* Norwell, MA: Kluwer Academic, 1991.

SAATY, THOMAS L., ED. *Lectures on Modern Mathematics,* 3 Vols. New York, NY: John Wiley, 1963–65.

SAWYER, W.W. *Introducing Mathematics,* 4 Vols. New York, NY: Penguin, 1964–70.

* SCHIFFER, M.M. AND BOWDEN, LEON. *The Role of Mathematics in Science.* Washington, DC: Mathematical Association of America, 1984.

** TIETZE, HEINRICH. *Famous Problems of Mathematics: Solved and Unsolved Mathematical Problems From Antiquity to Modern Times.* Baltimore, MD: Graylock Press, 1965.

1.4 Mathematics Appreciation Texts

* BAYLIS, JOHN AND HAGGARTY, ROD. *Alice in Numberland: A Students' Guide to the Enjoyment of Higher Mathematics.* Houndmills, England: Macmillan Education, 1988.

* BECK, ANATOLE; BLEICHER, MICHAEL N.; AND CROWE, DONALD W. *Excursions Into Mathematics.* New York, NY: Worth, 1969.

BILLSTEIN, RICK AND LOTT, JOHNNY W. *Mathematics for Liberal Arts: A Problem Solving Approach.* Redwood City, CA: Benjamin Cummings, 1986.

** CONSORTIUM FOR MATHEMATICS AND ITS APPLICATIONS. *For All Practical Purposes: Introduction to Contemporary Mathematics,* Second Edition. New York, NY: W.H. Freeman, 1988, 1991.

COUGHLIN, RAYMOND AND ZITARELLI, DAVID E. *The Ascent of Mathematics.* New York, NY: McGraw-Hill, 1984.

DAVIS, MORTON D. *Mathematically Speaking.* San Diego, CA: Harcourt Brace Jovanovich, 1980.

** DUNHAM, WILLIAM. *Journey Through Genius: The Great Theorems of Mathematics.* New York, NY: John Wiley, 1990.

** GARDINER, A. *Discovering Mathematics: The Art of Investigation.* New York, NY: Oxford University Press, 1987.

GOLOS, ELLERY B. *Patterns in Mathematics.* Boston, MA: Prindle, Weber and Schmidt, 1981.

GOWAR, NORMAN. *An Invitation to Mathematics.* New York, NY: Oxford University Press, 1979.

HERSTEIN, I.N. AND KAPLANSKY, IRVING. *Matters Mathematical,* Second Edition. New York, NY: Harper and Row, 1974; New York, NY: Chelsea, 1978.

* JACOBS, HAROLD R. *Mathematics: A Human Endeavor,* Second Edition. New York, NY: W.H. Freeman, 1970, 1982.

MESERVE, BRUCE E.; SOBEL, MAX A.; AND DOSSEY, JOHN A. *Contemporary Mathematics,* Fourth Edition. Englewood Cliffs, NJ: Prentice Hall, 1977, 1987.

MILLER, CHARLES D.; HEEREN, VERN E.; AND HORNSBY, E. JOHN, JR. *Mathematical Ideas,* Sixth Edition. Glenview, IL: Scott Foresman, 1970, 1990.

PENNEY, DAVID E. *Perspectives in Mathematics.* Reading, MA: W.A. Benjamin, 1972.

* RESNIKOFF, H.L. AND WELLS, R.O., JR. *Mathematics in Civilization.* New York, NY: Holt, Rinehart and Winston, 1973; Mineola, NY: Dover, 1984.

* ROBERTS, A. WAYNE AND VARBERG, DALE E. *Faces of Mathematics: An Introductory Course for College Students,* Second Edition. New York, NY: Thomas Y. Crowell, 1978; New York, NY: Harper and Row, 1982.

SMITH, KARL J. *Mathematics, Its Power and Utility,* Third Edition. Pacific Grove, CA: Brooks/Cole, 1983, 1989.

SMITH, KARL J. *The Nature of Mathematics,* Sixth Edition. Pacific Grove, CA: Brooks/Cole, 1991.

SONDHEIMER, ERNST AND ROGERSON, ALAN. *Numbers and Infinity: A Historical Account of Mathematical Concepts.* New York, NY: Cambridge University Press, 1981.

* STEIN, SHERMAN K. *Mathematics, The Man-made Universe: An Introduction to the Spirit of Mathematics,* Third Edition. New York, NY: W.H. Freeman, 1976.

WILLERDING, MARGARET F. AND ENGELSOHN, HAROLD S. *Mathematics: The Alphabet of Science,* Third Edition. New York, NY: John Wiley, 1977.

1.5 Fiction, Aphorisms, Epigrams

*** ABBOTT, EDWIN A. *Flatland.* Mineola, NY: Dover, 1952; Princeton, NJ: Princeton University Press, 1991.

** DEWDNEY, A.K. *The Planiverse.* New York, NY: Poseidon Press, 1984.

** EVES, HOWARD W. *Mathematical Circles,* Six Volume Series (Various Titles). Boston, MA: Prindle, Weber and Schmidt, 1969–88.

* FADIMAN, CLIFTON. *Fantasia Mathematica.* New York, NY: Simon and Schuster, 1958.

* FADIMAN, CLIFTON. *The Mathematical Magpie.* New York, NY: Simon and Schuster, 1962.

* KNUTH, DONALD E. *Surreal Numbers.* Reading, MA: Addison-Wesley, 1974.

MORITZ, ROBERT E. *On Mathematics.* (Former title: *Memorabilia Mathematica.*) Mineola, NY: Dover, 1914, 1958.

ROSE, NICHOLAS J. *Mathematical Maxims and Minims.* Raleigh, NC: Rome Press, 1988.

1.6 Miscellany

* BOLLOBÁS, BÉLA, ED. *Littlewood's Miscellany,* Revised Edition. New York, NY: Cambridge University Press, 1986.

* COHN, VICTOR. *News & Numbers.* Ames, IA: Iowa State University Press, 1989.

* DAVIS, PHILIP J. AND CHINN, WILLIAM G. *3.1416 And All That.* New York, NY: Birkhäuser, 1985.

* FOMENKO, ANATOLIĬ T. *Mathematical Impressions.* Providence, RI: American Mathematical Society, 1990.

* HUNTLEY, H.E. *The Divine Proportion: A Study in Mathematical Beauty.* Mineola, NY: Dover, 1970.

* LANG, SERGE. *The Beauty of Doing Mathematics: Three Public Dialogues.* New York, NY: Springer-Verlag, 1985.

MAULDIN, R. DANIEL. *The Scottish Book: Mathematics from the Scottish Café.* New York, NY: Birkhäuser, 1981.

PAGE, WARREN, ED. *Two-Year College Mathematics Readings.* Washington, DC: Mathematical Association of America, 1981.

RÉNYI, ALFRÉD. *Dialogues on Mathematics.* San Francisco, CA: Holden-Day, 1967.

STEWART, IAN AND JAWORSKI, JOHN, EDS. *Seven Years of Manifold, 1968–1980.* Nantwich, Cheshire: Shiva Publishing, 1981.

WECHSLER, JUDITH, ED. *On Aesthetics in Science.* Cambridge, MA: MIT Press, 1978.

2 Reference

2.1 Dictionaries

BALLENTYNE, D.W.G. AND LOVETT, D.R. *A Dictionary of Named Effects and Laws in Chemistry, Physics, and Mathematics,* Fourth Edition. New York, NY: Chapman and Hall, 1980.

BOROWSKI, E.J. AND BORWEIN, JONATHAN M. *Collins Reference Dictionary of Mathematics.* New York, NY: Harper Perennial 1991.

* DAINTITH, JOHN AND NELSON, R.D., EDS. *The Penguin Dictionary of Mathematics.* New York, NY: Penguin, 1989.

HOWSON, A. GEOFFREY. *A Handbook of Terms Used in Algebra and Analysis.* New York, NY: Cambridge University Press, 1972.

** JAMES, GLENN AND JAMES, ROBERT C., EDS. *Mathematical Dictionary,* Fourth Edition. New York, NY: Van Nostrand Reinhold, 1976.

SNEDDON, I.N., ED. *Encyclopedic Dictionary of Mathematics for Engineers and Applied Scientists.* Elmsford, NY: Pergamon Press, 1976.

TIETJEN, GARY L. *A Topical Dictionary of Statistics.* New York, NY: Chapman and Hall, 1986.

2.2 Handbooks

* ABRAMOWITZ, MILTON AND STEGUN, IRENE A., EDS. *Handbook of Mathematical Functions.* Mineola, NY: Dover, 1965.

BARNETT, STEPHEN AND CRONIN, T.M. *Mathematical Formulae for Engineering and Science Students,* Third Edition. London: Bradford University Press, 1979.

BARTSCH, HANS-JOCHEN AND LIEBSCHER, HERBERT. *Handbook of Mathematical Formulas.* New York, NY: Academic Press, 1974.

*** BEYER, WILLIAM H., ED. *CRC Handbook of Mathematical Sciences,* Sixth Edition. Boca Raton, FL: CRC Press, 1987.

* BEYER, WILLIAM H., ED. *CRC Standard Probability and Statistics Tables and Formulae.* Boca Raton, FL: CRC Press, 1990.

* BEYER, WILLIAM H., ED. *CRC Standard Mathematical Tables and Formulas,* 29th Edition. Boca Raton, FL: CRC Press, 1991.

** BOAS, RALPH P., JR., ED. *A.J. Lohwater's Russian-English Dictionary of the Mathematical Sciences,* Second Edition, Revised and Expanded. Providence, RI: American Mathematical Society, 1990.

BORWEIN, JONATHAN M. AND BORWEIN, PETER B. *A Dictionary of Real Numbers.* Belmont, CA: Wadsworth, 1990.

* BRONSHTEIN, I.N. AND SEMENDYAYEV, K.A. *Handbook of Mathematics.* New York, NY: Van Nostrand Reinhold, 1985.

** BURINGTON, RICHARD S. AND MAY, DONALD C., JR. *Handbook of Probability and Statistics with Tables,* Second Edition. New York, NY: McGraw-Hill, 1970.

* GOULD, SYDNEY H. *Russian for the Mathematician.* New York, NY: Springer-Verlag, 1972.

GRADSHTEYN, I.S. AND RYZHIK, I.M. *Table of Integrals, Series, and Products,* New York, NY: Academic Press, 1966, 1980.

GRAZDA, EDWARD E., ED. *Handbook of Applied Mathematics,* Fourth Edition. Melbourne, FL: Robert E. Krieger, 1977.

HUTCHINSON, T.P. AND LAI, C.D. *Continuous Bivariate Distributions, Emphasizing Applications.* Adelaide, Australia: Rumsby Scientific, 1990.

JOLLEY, L.B.W. *Summation of Series,* Second Revised Edition. Mineola, NY: Dover, 1961.

LAWRENCE, J. DENNIS. *A Catalog of Special Plane Curves.* Mineola, NY: Dover, 1972.

PATEL, JAGDISH K.; KAPADIA, C.H.; AND OWEN, D.B. *Handbook of Statistical Distributions.* New York, NY: Marcel Dekker, 1976.

** SLOANE, N.J.A. *A Handbook of Integer Sequences.* New York, NY: Academic Press, 1973.

SPANIER, JEROME AND OLDHAM, KEITH B. *An Atlas of Functions.* New York, NY: Hemisphere, 1987.

* VON SEGGERN, DAVID H. *CRC Handbook of Mathematical Curves and Surfaces.* Boca Raton, FL: CRC Press, 1990.

WELLS, DAVID. *The Penguin Dictionary of Curious and Interesting Geometry.* New York, NY: Penguin, 1991.

** ZWILLINGER, DANIEL. *Handbook of Differential Equations.* New York, NY: Academic Press, 1989.

2.3 Encyclopedias

ABBOTT, DAVID, ED. *The Biographical Dictionary of Scientists and Mathematicians.* New York, NY: Peter Bedrick, 1986.

*** AMERICAN COUNCIL OF LEARNED SOCIETIES. *Biographical Dictionary of Mathematicians.* New York, NY: Charles Scribner's, 1991.

* GELLERT, WALTER, *et al.*, EDS. *The VNR Concise Encyclopedia of Mathematics.* New York, NY: Van Nostrand Reinhold, 1975.

** GILLISPIE, CHARLES C., ED. *Dictionary of Scientific Biography,* 16 Vols. plus Suppl. New York, NY: Charles Scribner's, 1970–90.

*** IYANAGA, SHÔKICHI AND KAWADA, YUKIYOSI, EDS. *Encyclopedic Dictionary of Mathematics.* Cambridge, MA: MIT Press, 1977, 1986.

** KOTZ, SAMUEL AND JOHNSON, NORMAN L., EDS. *Encyclopedia of Statistical Sciences.* New York, NY: John Wiley, 1982–88.

KRUSKAL, WILLIAM H. AND TANUR, JUDITH M., EDS. *International Encyclopedia of Statistics.* New York, NY: Free Press, 1978.

SHAPIRO, MAX S., ED. *Mathematics Encyclopedia.* New York, NY: Doubleday, 1977.

2.4 Bibliographies

ANDERSON, T.W.; GUPTA, SOMESH DAS; AND STYAN, GEORGE P.H. *A Bibliography of Multivariate Statistical Analysis.* Melbourne, FL: Robert E. Krieger, 1977.

** DAUBEN, JOSEPH W. *The History of Mathematics from Antiquity to the Present: A Selective Bibliography.* New York, NY: Garland, 1985.

DICK, ELIE M. *Current Information Sources in Mathematics: An Annotated Guide to Books and Periodicals, 1960–1972.* Littleton, CO: Libraries Unlimited, 1973.

DORLING, A.R., ED. *Use of Mathematical Literature.* Woburn, MA: Butterworth, 1977.

* GAFFNEY, MATTHEW P. AND STEEN, LYNN ARTHUR. *Annotated Bibliography of Expository Writing in the Mathematical Sciences.* Washington, DC: Mathematical Association of America, 1976.

GREENWOOD, J. ARTHUR AND HARTLEY, H.O. *Guide to Tables in Mathematical Statistics.* Princeton, NJ: Princeton University Press, 1962.

HØYRUP, ELSE. *Books About Mathematics: A Bibliography.* Roskilde: Roskilde University Center, 1979.

KENDALL, MAURICE AND DOIG, A.G. *Bibliography of Statistical Literature.* New York, NY: Hafner Press, 1965.

LANCASTER, H.O. *Bibliography of Statistical Bibliographies.* Edinburgh, Scotland: Oliver and Boyd, 1968.

* MAY, KENNETH O. *Bibliography and Research Manual of the History of Mathematics.* Toronto: University of Toronto Press, 1973.

PEMBERTON, JOHN E. *How To Find Out in Mathematics,* Second Revised Edition. Elmsford, NY: Pergamon Press, 1970.

** SCHAAF, WILLIAM L. *A Bibliography of Recreational Mathematics,* 4 Vols. Reston, VA: National Council of Teachers of Mathematics, 1955–78.

SCHAAF, WILLIAM L. *Mathematics and Science: An Adventure in Postage Stamps.* Reston, VA: National Council of Teachers of Mathematics, 1978.

SCHAAF, WILLIAM L. *The High School Mathematics Library*. Reston, VA: National Council of Teachers of Mathematics, 1976, 1987.

SCHAEFER, BARBARA K. *Using the Mathematical Literature: A Practical Guide*. New York, NY: Marcel Dekker, 1979.

* STEEN, LYNN ARTHUR, ED. *Library Recommendations for Undergraduate Mathematics*. Washington, DC: Mathematical Association of America, 1992.

WHEELER, MARGARIETE M. *Mathematics Library: Elementary and Junior High School*, Fifth Edition. Reston, VA: National Council of Teachers of Mathematics, 1960, 1986.

2.5 Indexes

GREENWOOD, J. ARTHUR; OLKIN, INGRAM; AND SAVAGE, I. RICHARD. *Index to Volumes 1–31, 1930–1960: The Annals of Mathematical Statistics*. Hayward, CA: Institute of Mathematical Statistics, 1962.

* MAY, KENNETH O. *Index of the American Mathematical Monthly, Volumes 1–80 (1894–1973)*. Washington, DC: Mathematical Association of America, 1977.

NATIONAL COUNCIL OF TEACHERS OF MATHEMATICS. *Cumulative Index, The Arithmetic Teacher: 1974–1983*, Vols. 21–30. Reston, VA: National Council of Teachers of Mathematics, 1984.

* NATIONAL COUNCIL OF TEACHERS OF MATHEMATICS. *The Mathematics Teacher: Cumulative Indices*, Vols 1–68, 59–68, 69–78; 1908–65, 1966–75, 1976–85. Reston, VA: National Council of Teachers of Mathematics, 1967, 1976, 1988.

ROSS, IAN C. AND TUKEY, JOHN W. *Index to Statistics and Probability: Permuted Titles*. Los Altos, CA: R and D Press, 1975.

* SEEBACH, J. ARTHUR, JR. AND STEEN, LYNN ARTHUR, EDS. *Mathematics Magazine: 50 Year Index*. Washington, DC: Mathematical Association of America, 1978.

2.6 Reviews

BAUMSLAG, GILBERT, ED. *Reviews on Infinite Groups, 1940–1970*, 2 Vols. Providence, RI: American Mathematical Society, 1974.

BROWN, WILLIAM G., ED. *Reviews in Graph Theory, 1940–1978*, 4 Vols. Providence, RI: American Mathematical Society, 1980.

GOLUB, GENE H., ED. *Reviews in Numerical Analysis, 1980–1986*, 5 Vols. Providence, RI: American Mathematical Society, 1987.

GORENSTEIN, DANIEL, ED. *Reviews on Finite Groups, 1940–1970*. Providence, RI: American Mathematical Society, 1984.

GUY, RICHARD K., ED. *Reviews in Number Theory, 1973–1983*, 6 Vols. Providence, RI: American Mathematical Society, 1984.

HALMOS, PAUL R., ED. *Reviews in Operator Theory, 1980–1986*, 4 Vols. Providence, RI: American Mathematical Society, 1989.

JOHNSON, WILLIAM B., ED. *Reviews in Functional Analysis, 1980–1986*, 4 Vols. Providence, RI: American Mathematical Society, 1989.

KOHN, J.J. AND KRA, IRWIN, EDS. *Reviews in Complex Analysis, 1980–1986*, 4 Vols. Providence, RI: American Mathematical Society, 1989.

LEVEQUE, WILLIAM J., ED. *Reviews in Number Theory, 1940–1972*, 6 Vols. Providence, RI: American Mathematical Society, 1974.

MAGURN, BRUCE A., ED. *Reviews in K-Theory, 1940–1984*. Providence, RI: American Mathematical Society, 1985.

PROTTER, MURRAY H., ED. *Reviews in Partial Differential Equations, 1980–1986*, 5 Vols. Providence, RI: American Mathematical Society, 1988.

SMALL, LANCE W., ED. *Reviews in Ring Theory, 1960–1984*, 3 Vols. Providence, RI: American Mathematical Society, 1981, 1986.

TROMBA, ANTHONY J., ED. *Reviews in Global Analysis, 1980–1986*, 5 Vols. Providence, RI: American Mathematical Society, 1988.

2.7 International Congresses

GLEASON, ANDREW M., ED. *Proceedings of the International Congress of Mathematicians, August 3–11, 1986.* Providence, RI: American Mathematical Society, 1987.

JAMES, R.D., ED. *Proceedings of the International Congress of Mathematicians*, 2 Vols. Montreal: Canadian Mathematical Congress, 1975.

LEHTO, OLLI, ED. *Proceedings of the International Congress of Mathematicians, Helsinki, 1978*, 2 Vols. Helsinki: University of Helsinki, 1979.

OLECH, CZESLAW AND CIESIELSKI, ZBIGNIEW, EDS. *Proceedings of the International Congress of Mathematicians, August 16–24, 1983, Warszawa.* New York, NY: Elsevier Science, 1984.

3 History

3.1 Surveys

BELL, ERIC T. *Development of Mathematics,* Second Edition. New York, NY: McGraw-Hill, 1945.

*** BOYER, CARL B. AND MERZBACH, UTA C. *A History of Mathematics,* Second Edition. Princeton, NJ: Princeton University Press, 1985; New York, NY: John Wiley, 1968, 1991.

* BURTON, DAVID M. *The History of Mathematics: An Introduction,* Second Edition. Needham Heights, MA: Allyn and Bacon, 1985; Dubuque, IA: William C. Brown, 1991.

** CAJORI, FLORIAN. *A History of Mathematical Notations.* Peru, IL: Open Court, 1974.

* CAJORI, FLORIAN. *A History of Mathematics,* Fifth Edition. New York, NY: Chelsea, 1980, 1991.

COOLIDGE, JULIAN L. *The Mathematics of Great Amateurs,* Second Edition. New York, NY: Oxford University Press, 1949, 1990.

DEDRON, P. AND ITARD, J. *Mathematics and Mathematicians,* 2 Vols. London: Transworld, 1973.

DIEUDONNÉ, JEAN. *Abrégé d'histoire des Mathématiques, 1700–1900,* 2 Vols. Paris: Hermann, 1978.

*** EVES, HOWARD W. *An Introduction to the History of Mathematics with Cultural Connections,* Sixth Edition. New York, NY: Rinehart and Co., 1953; Philadelphia, PA: Saunders College, 1990.

*** FAUVEL, JOHN AND GRAY, JEREMY, EDS. *The History of Mathematics: A Reader.* London: Macmillan Press and Open University, 1987.

HOLLINGDALE, STUART. *Makers of Mathematics.* New York, NY: Penguin, 1989.

*** KLINE, MORRIS. *Mathematics in Western Culture.* New York, NY: Oxford University Press, 1953, 1971.

*** KLINE, MORRIS. *Mathematical Thought from Ancient to Modern Times.* New York, NY: Oxford University Press, 1972.

** KRAMER, EDNA E. *The Nature and Growth of Modern Mathematics.* New York, NY: Fawcett World Library, 1974; Princeton, NJ: Princeton University Press, 1982.

MORGAN, BRYAN. *Men and Discoveries in Mathematics.* Central Islip, NY: Transatlantic Arts, 1972.

PHILLIPS, ESTHER R., ED. *Studies in the History of Mathematics.* Washington, DC: Mathematical Association of America, 1987.

ROWE, DAVID E. AND MCCLEARY, JOHN, EDS. *The History of Modern Mathematics,* 2 Vols. New York, NY: Academic Press, 1989.

* SMITH, DAVID EUGENE. *History of Mathematics,* 2 Vols. New York, NY: Ginn and Co., 1923; Mineola, NY: Dover, 1958.

* STILLWELL, JOHN. *Mathematics and Its History.* New York, NY: Springer-Verlag, 1989.

*** STRUIK, DIRK JAN. *A Concise History of Mathematics,* Fourth Revised Edition. Mineola, NY: Dover, 1987.

TEMPLE, GEORGE. *100 Years of Mathematics: A Personal Viewpoint.* New York, NY: Springer-Verlag, 1981.

3.2 Biographies

AITON, E.J. *Leibniz: A Biography.* New York, NY: Adam Hilger, 1985.

*** ALBERS, DONALD J. AND ALEXANDERSON, GERALD L., EDS. *Mathematical People: Profiles and Interviews.* New York, NY: Birkhäuser, 1985.

*** ALBERS, DONALD J., *et al.*, EDS. *More Mathematical People.* New York, NY: Academic Press, 1990.

ASPRAY, WILLIAM. *John von Neumann and the Origins of Modern Computing.* Cambridge, MA: MIT Press, 1990.

BAUM, JOAN. *The Calculating Passion of Ada Byron.* Hamden, CT: Archon Books, 1986.

* BEDINI, SILVIO. *The Life of Benjamin Banneker.* New York, NY: Charles Scribner's, 1972.

* BELHOSTE, BRUNO. *Augustin-Louis Cauchy: A Biography.* New York, NY: Springer-Verlag, 1991.

** BELL, ERIC T. *Men of Mathematics.* New York, NY: Simon and Schuster, 1937.

* BOX, JOAN FISHER. *R.A. Fisher: The Life of a Scientist.* New York, NY: John Wiley, 1978.

* BREWER, JAMES W. AND SMITH, MARTHA K., EDS. *Emmy Noether: A Tribute to Her Life and Work.* New York, NY: Marcel Dekker, 1981.

BUCCIARELLI, LOUIS L. AND DWORSKY, NANCY. *Sophie Germain: An Essay in the History of the Theory of Elasticity.* Norwell, MA: D. Reidel, 1980.

** BÜHLER, WALTER K. *Gauss: A Biographical Study.* New York, NY: Springer-Verlag, 1981.

BURKS, ALICE R. AND BURKS, ARTHUR W. *The First Electronic Computer: The Atanasoff Story.* Ann Arbor, MI: University of Michigan Press, 1988.

CALINGER, RONALD. *Gottfried Wilhelm Leibniz.* Troy, NY: Rensselaer Polytech Institute, 1976.

* COOKE, ROGER. *The Mathematics of Sonya Kovalevskaya.* New York, NY: Springer-Verlag, 1984.

COOPER, NECIA G., ED. *From Cardinals to Chaos: Reflections on the Life and Legacy of Stanislaw Ulam.* New York, NY: Cambridge University Press, 1989.

* DAUBEN, JOSEPH W. *Georg Cantor: His Mathematics and Philosophy of the Infinite.* Cambridge, MA: Harvard University Press, 1979; Princeton, NJ: Princeton University Press, 1990.

** DICK, AUGUSTE. *Emmy Noether, 1882–1935.* New York, NY: Birkhäuser, 1981.

* FAUVEL, JOHN, *et al.*, EDS. *Let Newton Be! A New Perspective on His Life and Works.* New York, NY: Oxford University Press, 1988.

GJERTSEN, DEREK. *The Newton Handbook.* London: Routledge and Kegan Paul, 1986.

GRATTAN-GUINNESS, I. *Joseph Fourier, 1768–1830.* Cambridge, MA: MIT Press, 1972.

* HALMOS, PAUL R. *I Want To Be A Mathematician: An Automathography in Three Parts.* Washington, DC: Mathematical Association of America, 1985; New York, NY: Springer-Verlag, 1985.

HANKINS, THOMAS L. *Sir William Rowan Hamilton.* Baltimore, MD: Johns Hopkins University Press, 1980.

* HARDY, G.H. *Ramanujan,* Third Edition. New York, NY: Chelsea, 1960, 1968.

HEYDE, C.C. AND SENETA, E. *I.J. Bienaymé: Statistical Theory Anticipated.* New York, NY: Springer-Verlag, 1977.

** HODGES, ANDREW. *Alan Turing, The Enigma.* New York, NY: Simon and Schuster, 1983.

* HOFFMANN, BANESH. *Albert Einstein: Creator and Rebel.* New York, NY: Viking Press, 1972.

* HYMAN, ANTHONY. *Charles Babbage: Pioneer of the Computer.* Princeton, NJ: Princeton University Press, 1982.

INFELD, LEOPOLD. *Whom the Gods Love: The Story of Evariste Galois.* Reston, VA: National Council of Teachers of Mathematics, 1978.

* KAC, MARK. *Enigmas of Chance: An Autobiography.* New York, NY: Harper and Row, 1985.

** KANIGEL, ROBERT. *The Man Who Knew Infinity: A Life of the Indian Genius Ramanujan.* New York, NY: Charles Scribner's, 1991.

KENNEDY, HUBERT C. *Peano: Life and Works of Giuseppe Peano.* Norwell, MA: D. Reidel, 1980.

** KOBLITZ, ANN HIBNER. *A Convergence of Lives, Sofia Kovalevskaia: Scientist, Writer, Revolutionary.* New York, NY: Birkhäuser, 1983.

* LÜTZEN, JESPER. *Joseph Liouville, 1805–1882: Master of Pure and Applied Mathematics.* New York, NY: Springer-Verlag, 1990.

* MACHALE, DESMOND. *George Boole: His Life and Work.* Dublin: Boole Press, 1985.

MAHONEY, MICHAEL S. *The Mathematical Career of Pierre de Fermat, 1601–1665.* Princeton, NJ: Princeton University Press, 1973.

* MASANI, PESI. *Norbert Wiener, 1894–1964.* New York, NY: Birkhäuser, 1990.

* O'DONNELL, SEAN. *William Rowan Hamilton: Portrait of a Prodigy.* Dublin: Boole Press, 1983.

** ORE, OYSTEIN. *Cardano: The Gambling Scholar.* Princeton, NJ: Princeton University Press, 1953; Mineola, NY: Dover, 1965.

** ORE, OYSTEIN. *Niels Henrik Abel: Mathematician Extraordinary.* Minneapolis, MN: University of Minnesota Press, 1957; New York, NY: Chelsea, 1974.

* PARIKH, CAROL. *The Unreal Life of Oscar Zariski.* San Diego, CA: Academic Press, 1991.

** REID, CONSTANCE. *Courant in Göttingen and New York.* New York, NY: Springer-Verlag, 1976.

*** REID, CONSTANCE. *Hilbert.* (Combined edition, *Hilbert-Courant.*) New York, NY: Springer-Verlag, 1970, 1986.

** REID, CONSTANCE. *Neyman—From Life.* New York, NY: Springer-Verlag, 1982.

RUSSELL, BERTRAND. *The Autobiography of Bertrand Russell,* 2 Vols. New York, NY: Simon and Schuster, 1968, 1969.

SCOTT, J.F. *The Mathematical Work of John Wallis,* Second Edition. New York, NY: Chelsea, 1981.

SCOTT, J.F. *The Scientific Work of René Descartes (1596–1650).* Bristol, PA: Taylor and Francis, 1952, 1976.

SRINIVASAN, BHAMA AND SALLY, JUDITH D., EDS. *Emmy Noether in Bryn Mawr.* New York, NY: Springer-Verlag, 1983.

** ULAM, S.M. *Adventures of a Mathematician.* New York, NY: Charles Scribner's, 1976.

VAN STIGT, WALTER P. *Brouwer's Intuitionism.* Amsterdam: North-Holland, 1990.

WANG, HAO. *Reflections on Kurt Gödel.* Cambridge, MA: MIT Press, 1987.

** WESTFALL, RICHARD S. *Never at Rest: A Biography of Isaac Newton.* New York, NY: Cambridge University Press, 1983.

WHITEMORE, HUGH. *Breaking the Code.* Oxford: Amber Lane Press, 1987.

* WIENER, NORBERT. *Ex-Prodigy: My Childhood and Youth* and *I Am A Mathematician: Life of a Prodigy.* Cambridge, MA: MIT Press, 1953, 1956.

3.3 Source Books

** BIRKHOFF, GARRETT, ED. *A Source Book in Classical Analysis.* Cambridge, MA: Harvard University Press, 1973.

* CALINGER, RONALD, ED. *Classics of Mathematics.* Oak Park, IL: Moore, 1982.

GRANT, EDWARD, ED. *A Source Book in Medieval Science.* Cambridge, MA: Harvard University Press, 1974.

** SMITH, DAVID EUGENE. *A Source Book in Mathematics.* New York, NY: McGraw-Hill, 1929; Mineola, NY: Dover, 1959.

** STRUIK, DIRK JAN, ED. *A Source Book in Mathematics, 1200–1800.* Cambridge, MA: Harvard University Press, 1969; Princeton, NJ: Princeton University Press, 1986.

* VAN HEIJENOORT, JEAN, ED. *From Frege to Gödel: A Source Book in Mathematical Logic, 1879–1931.* Cambridge, MA: Harvard University Press, 1967.

3.4 Classic Works

BERNDT, BRUCE C. *Ramanujan's Notebooks,* Parts I-III. New York, NY: Springer-Verlag, 1985–91.

BOOLE, GEORGE. *Treatise on the Calculus of Finite Differences,* Fifth Edition. New York, NY: Chelsea, 1958.

* CARDANO, GIROLAMO. *The Great Art or the Rules of Algebra.* Cambridge, MA: MIT Press, 1968.

* CHACE, ARNOLD B. *The Rhind Mathematical Papyrus.* Reston, VA: National Council of Teachers of Mathematics, 1979.

CHRYSTAL, GEORGE. *Textbook of Algebra,* 2 Vols., Seventh Edition. New York, NY: Chelsea, 1964.

DE MOIVRE, ABRAHAM. *The Doctrine of Chances or a Method of Calculating the Probabilities of Events in Play,* Third Edition. New York, NY: Chelsea, 1967.

** DEDEKIND, RICHARD. *Essays on the Theory of Numbers.* Mineola, NY: Dover, 1963.

DEE, JOHN. *The Mathematicall Praeface to the Elements of Geometrie of Euclid of Megara (1570).* New York, NY: Science History, 1975.

** DIJKSTERHUIS, E.J. *Archimedes.* Princeton, NJ: Princeton University Press, 1987.

* EULER, LEONHARD. *Elements of Algebra.* New York, NY: Springer-Verlag, 1984.

* EULER, LEONHARD. *Introduction to Analysis of the Infinite,* 2 Vols. New York, NY: Springer-Verlag, 1988, 1990.

FEDERICO, P.J. *Descartes on Polyhedra: A Study of the De Solidorum Elementis.* New York, NY: Springer-Verlag, 1982.

FLEGG, H. GRAHAM; HAY, CYNTHIA; AND MOSS, BARBARA, EDS. *Nicolas Chuquet: Renaissance Mathematician.* Norwell, MA: D. Reidel, 1985.

GALILEI, GALILEO. *Dialogues Concerning Two New Sciences.* Mineola, NY: Dover, 1954.

* GAUSS, CARL FRIEDRICH. *Disquisitiones Arithmeticae,* English Edition. New York, NY: Springer-Verlag, 1986.

HALSTED, GEORGE B., ED. *Girolamo Saccheri's Euclides Vindicatus.* New York, NY: Chelsea, 1986.

** HEATH, THOMAS L., ED. *The Works of Archimedes.* New York, NY: Cambridge University Press, 1897; Mineola, NY: Dover, 1953, 1964.

* HEATH, THOMAS L. *Apollonius of Perga.* Cambridge: W. Haffer, 1986.

HEATH, THOMAS L. *Aristarchus of Samos: The Ancient Copernicus.* Oxford: Clarendon Press, 1913; Mineola, NY: Dover, 1981.

* HEATH, THOMAS L. *Diophantus of Alexandria.* Cambridge: Cambridge University Press, 1885; Mineola, NY: Dover, 1964.

*** HEATH, THOMAS L. *The Thirteen Books of Euclid's Elements,* 3 Vols. Mineola, NY: Dover, 1956.

HUGHES, BARNABAS, ED. *Jordanus de Nemore: De numeris datis.* Berkeley, CA: University of California Press, 1981.

LAM, LAY YONG. *A Critical Study of the Yang Hui Suan Fa: A Thirteenth-century Chinese Mathematical Treatise.* Singapore: Singapore University Press, 1977.

MOHR, GEORG. *Compendium Euclidis Curiosi: Or Geometrical Operations.* Copenhagen: Georg Mohr Foundation, 1982.

MÜLLER, JOHANNES. *Regiomontanus: On Triangles.* Madison, WI: University of Wisconsin Press, 1967.

NAPIER, JOHN. *Rabdology.* Cambridge, MA: MIT Press, 1990.

*** NEWTON, ISAAC. *Mathematical Principles of Natural Philosophy.* Berkeley, CA: University of California Press, 1934.

* RAMANUJAN, SRINIVASA. *Collected Papers.* New York, NY: Cambridge University Press, 1927; New York, NY: Chelsea, 1962.

* ROBINS, GAY AND SHUTE, CHARLES. *The Rhind Mathematical Papyrus: An Ancient Egyptian Text.* London: British Museum, 1987.

SAIDAN, A.S. *The Arithmetic of Al-Uqlīdisī.* Norwell, MA: D. Reidel, 1978.

SIGLER, L.E. *Leonardo Pisano Fibonacci: The Book of Squares.* New York, NY: Academic Press, 1987.

* SMITH, DAVID EUGENE AND LATHAM, MARCIA L. *The Geometry of René Descartes.* Mineola, NY: Dover, 1954.

SWERDLOW, N.M. AND NEUGEBAUER, O. *Mathematical Astronomy in Copernicus's De Revolutionibus.* New York, NY: Springer-Verlag, 1984.

* SWETZ, FRANK J. *Capitalism and Arithmetic: The New Math of the 15th Century,* incl. *Treviso Arithmetic* (1487). La Salle, IL: Open Court, 1987.

* TOOMER, G.J. *Ptolemy's Almagest.* New York, NY: Springer-Verlag, 1984.

VIÈTE, FRANÇOIS. *The Analytic Art.* Kent, OH: Kent State University Press, 1983.

WHITESIDE, D.T., ED. *The Mathematical Papers of Isaac Newton,* 8 Vols. New York, NY: Cambridge University Press, 1967–81.

YOUNG, WILLIAM H. AND YOUNG, GRACE CHISHOLM. *The Theory of Sets of Points,* Second Edition. New York, NY: Chelsea, 1972.

3.5 Ancient and Medieval

** AABOE, ASGER. *Episodes from the Early History of Mathematics.* Washington, DC: Mathematical Association of America, 1964.

ALLMAN, GEORGE J. *Greek Geometry from Thales to Euclid.* New York, NY: Arno Press, 1976.

** BERGGREN, J.L. *Episodes in the Mathematics of Medieval Islam.* New York, NY: Springer-Verlag, 1986.

* EDWARDS, A.W.F. *Pascal's Arithmetical Triangle.* New York, NY: Oxford University Press, 1987.

* EVES, HOWARD W. *Great Moments in Mathematics (Before 1650).* Washington, DC: Mathematical Association of America, 1980.

GILLINGS, RICHARD J. *Mathematics in the Times of the Pharaohs.* Mineola, NY: Dover, 1982.

*** HEATH, THOMAS L. *A History of Greek Mathematics,* 2 Vols. Mineola, NY: Dover, 1981.

JOSEPH, GEORGE GHEVARGESE. *The Crest of a Peacock: The Non-European Roots of Mathematics.* New York, NY: Penguin, 1990.

KLEIN, JACOB. *Greek Mathematical Thought and the Origin of Algebra.* Cambridge, MA: MIT Press, 1968.

KNORR, WILBUR R. *The Ancient Tradition of Geometric Problems.* New York, NY: Birkhäuser, 1986.

** LĬ YĂN AND DÙ SHÍRÁN. *Chinese Mathematics: A Concise History.* New York, NY: Clarendon Press, 1987.

MIKAMI, YASHIO. *The Development of Mathematics in China and Japan,* Second Edition. New York, NY: Chelsea, 1974.

MUELLER, IAN. *Philosophy of Mathematics and Deductive Structure in Euclid's Elements.* Cambridge, MA: MIT Press, 1981.

NEUGEBAUER, O. *A History of Ancient Mathematical Astronomy,* 3 Vols. New York, NY: Springer-Verlag, 1975.

* NEUGEBAUER, O. *The Exact Sciences in Antiquity,* Second Edition. Providence, RI: Brown University Press, 1970.

* SWETZ, FRANK J. AND KAO, T.I. *Was Pythagoras Chinese? An Examination of Right Triangle Theory in Ancient China.* Reston, VA: National Council of Teachers of Mathematics, 1977.

VAN DER WAERDEN, B.L. *Geometry and Algebra in Ancient Civilizations.* New York, NY: Springer-Verlag, 1983.

** VAN DER WAERDEN, B.L. *Science Awakening,* 2 Vols. Groningen: Wolters-Noordhoff, 1954; New York, NY: Oxford University Press, 1961; Princeton Junction, NJ: Scholar's Bookshelf, 1988.

** WILDER, RAYMOND L. *Evolution of Mathematical Concepts: An Elementary Study.* New York, NY: John Wiley, 1968; New York, NY: Halsted Press, 1973.

ZINNER, ERNST. *Regiomontanus: His Life and Work.* Amsterdam: North-Holland, 1990.

3.6 Modern

ALBERS, DONALD J.; ALEXANDERSON, GERALD L.; AND REID, CONSTANCE. *International Mathematical Congresses: An Illustrated History, 1893–1986.* New York, NY: Springer-Verlag, 1987.

DAUBEN, JOSEPH W., ED. *Mathematical Perspectives: Essays on Mathematics and Its Historical Development.* New York, NY: Academic Press, 1981.

** DUREN, PETER; ASKEY, RICHARD A.; AND MERZBACH, UTA C., EDS. *A Century of Mathematics in America,* 3 Vols. Providence, RI: American Mathematical Society, 1988–1989.

* EVES, HOWARD W. *Great Moments in Mathematics (After 1650).* Washington, DC: Mathematical Association of America, 1981.

HALMOS, PAUL R. *I Have A Photographic Memory.* Providence, RI: American Mathematical Society, 1987.

PÓLYA, GEORGE. *The Pólya Picture Album.* New York, NY: Birkhäuser, 1987.

TARWATER, J. DALTON, *et al.*, EDS. *American Mathematical Heritage: Algebra and Applied Mathematics.* Lubbock, TX: Texas Technical University Press, 1981.

TARWATER, J. DALTON; WHITE, JOHN T.; AND MILLER, JOHN D., EDS. *Men and Institutions in American Mathematics.* Lubbock, TX: Texas Technical University Press, 1976.

3.7 Numbers

BECKMANN, PETR. *A History of π(pi),* Third Edition. Boulder, CO: Golem Press, 1970, 1974.

CROSSLEY, J.N. *The Emergence of Number.* Teaneck, NJ: World Scientific, 1987.

CRUMP, THOMAS. *The Anthropology of Numbers.* New York, NY: Cambridge University Press, 1990.

*** DICKSON, LEONARD E. *History of the Theory of Numbers,* 3 Vols. New York, NY: Chelsea, 1952, 1971.

FLEGG, H. GRAHAM, ED. *Numbers Through the Ages.* New York, NY: Macmillan, 1989.

* IFRAH, GEORGES. *From One to Zero: A Universal History of Numbers.* New York, NY: Viking Press, 1985.

KARPINSKI, LOUIS C. *The History of Arithmetic.* Chicago, IL: Rand McNally, 1925; New York, NY: Russell and Russell, 1965.

* MENNINGER, KARL. *Number Words and Number Symbols: A Cultural History of Numbers.* Cambridge, MA: MIT Press, 1977.

3.8 Algebra

CHANDLER, BRUCE AND MAGNUS, WILHELM. *The History of Combinatorial Group Theory: A Case Study in the History of Ideas.* New York, NY: Springer-Verlag, 1982.

NOVÝ, LUBOŠ. *Origins of Modern Algebra.* Groningen: Wolters-Noordhoff, 1973.

** VAN DER WAERDEN, B.L. *A History of Algebra: From al-Khwārizmī to Emmy Noether.* New York, NY: Springer-Verlag, 1985.

* WUSSING, HANS. *The Genesis of the Abstract Group Concept.* Cambridge, MA: MIT Press, 1984.

3.9 Calculus and Analysis

ARNOLD, V.I. *Huygens and Barrow, Newton and Hooke.* New York, NY: Birkhäuser, 1990.

* BARON, MARGARET E. *The Origins of the Infinitesimal Calculus.* Mineola, NY: Dover, 1987.

* BOTTAZZINI, UMBERTO. *The Higher Calculus: A History of Real and Complex Analysis from Euler to Weierstrass.* New York, NY: Springer-Verlag, 1986.

** BOYER, CARL B. *The History of Calculus and Its Conceptual Development.* Mineola, NY: Dover, 1959.

CROWE, MICHAEL J. *A History of Vector Analysis: The Evolution of the Idea of a Vectorial System.* Notre Dame, IN: University of Notre Dame Press, 1967; Mineola, NY: Dover, 1985.

DIEUDONNÉ, JEAN. *History of Functional Analysis.* New York, NY: Elsevier Science, 1981.

** EDWARDS, C.H., JR. *The Historical Development of the Calculus.* New York, NY: Springer-Verlag, 1979.

ENGELSMAN, STEVEN B. *Families of Curves and the Origins of Partial Differentiation.* New York, NY: Elsevier Science, 1984.

GOLDSTINE, HERMAN H. *A History of Numerical Analysis from the 16th Through the 19th Century.* New York, NY: Springer-Verlag, 1977.

GOLDSTINE, HERMAN H. *A History of the Calculus of Variations from the 17th Through the 19th Century.* New York, NY: Springer-Verlag, 1980.

** GRABINER, JUDITH V. *The Origins of Cauchy's Rigorous Calculus.* Cambridge, MA: MIT Press, 1981.

* GRATTAN-GUINNESS, I., ED. *From the Calculus to Set Theory, 1630–1910: An Introductory History.* London: Gerald Duckworth, 1980.

GUICCIARDINI, NICCOLÒ. *The Development of Newtonian Calculus in Britain, 1700–1800.* New York, NY: Cambridge University Press, 1989.

HALL, A. RUPERT. *Philosophers at War: The Quarrel Between Newton and Leibniz.* New York, NY: Cambridge University Press, 1980.

* HAWKINS, THOMAS. *Lebesgue's Theory of Integration: Its Origins and Development,* Second Edition. Madison, WI: University of Wisconsin Press, 1970; New York, NY: Chelsea, 1975.

LEBESGUE, HENRI. *Measure and the Integral.* San Francisco, CA: Holden-Day, 1966.

MANHEIM, JEROME H. *The Genesis of Point Set Topology.* Elmsford, NY: Pergamon Press, 1964.

MONNA, A.F. *Functional Analysis in Historical Perspective.* New York, NY: Oosthoek, Scheltema and Holkema, 1973.

* MOORE, GREGORY H. *Zermelo's Axiom of Choice: Its Origins, Development, and Influence.* New York, NY: Springer-Verlag, 1982.

VAN DALEN, DIRK AND MONNA, A.F. *Sets and Integration: An Outline of the Development.* Groningen: Wolters-Noordhoff, 1972.

3.10 Geometry

* BONOLA, ROBERTO. *Non-Euclidean Geometry: A Critical and Historical Study of Its Development.* Mineola, NY: Dover, 1955.

* BOYER, CARL B. *History of Analytic Geometry.* New York, NY: Scripta Mathematica, 1956; Princeton Junction, NJ: Scholar's Bookshelf, 1988.

COOLIDGE, JULIAN L. *A History of Geometrical Methods.* Mineola, NY: Dover, 1963.

* COOLIDGE, JULIAN L. *A History of the Conic Sections and Quadric Surfaces.* Oxford: Oxford University Press, 1945; Mineola, NY: Dover, 1968.

DIEUDONNÉ, JEAN. *A History of Algebraic and Differential Topology, 1900–1960.* New York, NY: Birkhäuser, 1989.

DIEUDONNÉ, JEAN. *History of Algebraic Geometry.* Belmont, CA: Wadsworth, 1985.

KLEIN, FELIX. *Famous Problems of Elementary Geometry.* Boston, MA: Ginn and Co, 1897.

* ROSENFELD, B.A. *A History of Non-Euclidean Geometry: Evolution of the Concept of a Geometric Space.* New York, NY: Springer-Verlag, 1988.

YAGLOM, I.M. *Felix Klein and Sophus Lie: Evolution of the Idea of Symmetry in the Nineteenth Century.* New York, NY: Birkhäuser, 1988.

3.11 Probability and Statistics

* ADAMS, WILLIAM J. *The Life and Times of the Central Limit Theorem.* New York, NY: Kaedmon, 1974.

COWLES, MICHAEL. *Statistics in Psychology: An Historical Perspective.* Hillsdale, NJ: Lawrence Erlbaum, 1989.

DASTON, LORRAINE. *Classical Probability in the Enlightenment.* Princeton, NJ: Princeton University Press, 1988.

* GANI, J., ED. *The Making of Statisticians.* New York, NY: Springer-Verlag, 1982.

* GIGERENZER, GERD, *et al.* *The Empire of Chance: How Probability Changed Science and Everyday Life.* New York, NY: Cambridge University Press, 1989.

HACKING, IAN. *The Emergence of Probability.* New York, NY: Cambridge University Press, 1975.

HALD, ANDERS. *A History of Probability and Statistics and Their Applications Before 1750.* New York, NY: John Wiley, 1990.

KRÜGER, LORENZ, *et al.*, EDS. *The Probabilistic Revolution,* 2 Vols. Cambridge, MA: MIT Press, 1987.

* MAISTROV, LEONID E. *Probability Theory: A Historical Sketch.* New York, NY: Academic Press, 1974.

OWEN, D.B., ED. *On the History of Statistics and Probability.* New York, NY: Marcel Dekker, 1976.

PORTER, THEODORE M. *The Rise of Statistical Thinking, 1820–1900.* Princeton, NJ: Princeton University Press, 1986.

*** STIGLER, STEPHEN M. *The History of Statistics: The Measurement of Uncertainty Before 1900.* Cambridge, MA: Harvard University Press, 1986.

* TODHUNTER, I. *A History of the Mathematical Theory of Probability.* New York, NY: Chelsea, 1965.

WALKER, HELEN M. *Studies in the History of Statistical Method.* Baltimore, MD: Williams and Wilkins, 1929; New York, NY: Arno Press, 1975.

3.12 Computers

AUGARTEN, STAN. *Bit by Bit: An Illustrated History of Computers.* London: George Allen and Unwin, 1985.

EAMES, CHARLES AND EAMES, RAY. *A Computer Perspective.* Cambridge, MA: Harvard University Press, 1973.

FREIBERGER, PAUL AND SWAINE, MICHAEL. *Fire in the Valley: The Making of the Personal Computer.* Berkeley, CA: Osborne/McGraw-Hill, 1984.

* GOLDSTINE, HERMAN H. *The Computer from Pascal to von Neumann.* Princeton, NJ: Princeton University Press, 1972, 1980.

LINDGREN, MICHAEL. *Glory and Failure.* Cambridge, MA: MIT Press, 1990.

METROPOLIS, N.; HOWLETT, J.; AND ROTA, GIAN-CARLO, EDS. *A History of Computing in the Twentieth Century: A Collection of Essays.* New York, NY: Academic Press, 1980.

RANDELL, BRIAN. *The Origins of Digital Computers: Selected Papers.* New York, NY: Springer-Verlag, 1973.

STERN, NANCY. *From ENIAC to UNIVAC: An Appraisal of the Eckert-Mauchly Computers.* Bedford, MA: Digital Press, 1981.

3.13 Physics and Astronomy

MEHRA, JAGDISH. *Einstein, Hilbert, and The Theory of Gravitation: Historical Origins of General Relativity Theory.* Norwell, MA: D. Reidel, 1974.

TRUESDELL, C. *The Tragicomical History of Thermodynamics, 1822–1854.* New York, NY: Springer-Verlag, 1980.

YODER, JOELLA G. *Unrolling Time: Christiaan Huygens and the Mathematization of Nature.* New York, NY: Cambridge University Press, 1988.

4 Recreational Mathematics

4.1 Surveys

* AVERBACH, BONNIE AND CHEIN, ORIN. *Mathematics: Problem Solving through Recreational Mathematics.* New York, NY: W.H. Freeman, 1980.

*** BALL, W.W. ROUSE AND COXETER, H.S.M. *Mathematical Recreations and Essays,* Thirteenth Edition. Toronto: University of Toronto Press, 1974; Mineola, NY: Dover, 1987.

* BEILER, ALBERT. *Recreations in the Theory of Numbers.* Mineola, NY: Dover, 1964.

*** BERLEKAMP, ELWYN R.; CONWAY, JOHN HORTON; AND GUY, RICHARD K. *Winning Ways for Your Mathematical Plays,* 2 Vols. New York, NY: Academic Press, 1982.

CADWELL, J.H. *Topics in Recreational Mathematics.* New York, NY: Cambridge University Press, 1966.

FOSTER, JAMES E. *Mathematics as Diversion.* Astoria, IL: Fulton County Press, 1978.

*** KRAITCHIK, MAURICE. *Mathematical Recreations,* Second Edition. Mineola, NY: Dover, 1953.

* MELZAK, Z.A. *Companion to Concrete Mathematics,* 2 Vols. New York, NY: John Wiley, 1973, 1976.

OGILVY, C. STANLEY. *Tomorrow's Math: Unsolved Problems for the Amateur,* Second Edition. New York, NY: Oxford University Press, 1972.

PÉTER, RÓZSA. *Playing with Infinity: Mathematical Explorations and Excursions.* Mineola, NY: Dover, 1976.

SCHUH, FRED. *The Master Book of Mathematical Recreations.* Mineola, NY: Dover, 1968.

SPRAGUE, ROLAND. *Recreation in Mathematics.* Mineola, NY: Dover, 1963.

4.2 Games and Puzzles

ANDREWS, WILLIAM S. *Magic Squares and Cubes,* Second Edition. Mineola, NY: Dover, 1960.

APSIMON, HUGH. *Mathematical Byways in Ayling, Beeling, and Ceiling.* New York, NY: Oxford University Press, 1984.

* BEASLEY, JOHN D. *The Ins and Outs of Peg Solitaire.* New York, NY: Oxford University Press, 1985.

BELL, ROBBIE AND CORNELIUS, MICHAEL. *Board Games Round the World: A Resource Book for Mathematical Investigations.* New York, NY: Cambridge University Press, 1990.

BERRONDO, MARIE. *Mathematical Games.* Englewood Cliffs, NJ: Prentice Hall, 1983.

BRANDRETH, GYLES. *Numberplay.* New York, NY: Rawson, 1984.

** CARROLL, LEWIS. *Mathematical Recreations of Lewis Carroll,* 2 Vols. Mineola, NY: Dover, 1958.

DOMORYAD, A.P. *Mathematical Games and Pastimes.* Elmsford, NY: Pergamon Press, 1963.

EISS, HARRY EDWIN. *Dictionary of Mathematical Games, Puzzles, and Amusements.* Westport, CT: Greenwood Press, 1988.

* EWING, JOHN AND KOŚNIOWSKI, CZES. *Puzzle It Out: Cubes, Groups and Puzzles.* New York, NY: Cambridge University Press, 1982.

FREY, ALEXANDER H., JR. AND SINGMASTER, DAVID. *Handbook of Cubik Math.* Hillside, NJ: Enslow, 1982.

** HORDERN, EDWARD. *Sliding Piece Puzzles.* New York, NY: Oxford University Press, 1986.

LINES, MALCOLM. *Think of a Number.* New York, NY: Adam Hilger, 1990.

** O'BEIRNE, T.H. *Puzzles and Paradoxes: Fascinating Excursions in Recreational Mathematics.* Mineola, NY: Dover, 1984.

* READ, RONALD C. *Tangrams—330 Puzzles.* Mineola, NY: Dover, 1965.

RUBIK, ERNŐ, *et al. Rubik's Cubic Compendium.* New York, NY: Oxford University Press, 1987.

STEWART, IAN. *Game, Set, and Math: Enigmas and Conundrums.* Cambridge, MA: Basil Blackwell, 1989.

4.3 Puzzle Collections

AINLEY, STEPHEN. *Mathematical Puzzles.* Englewood Cliffs, NJ: Prentice Hall, 1983.

BARR, STEPHEN. *Mathematical Brain Benders: 2nd Miscellany of Puzzles.* Mineola, NY: Dover, 1982.

BERLOQUIN, PIERRE. *The Garden of the Sphinx: 150 Challenging and Instructive Puzzles.* New York, NY: Charles Scribner's, 1985.

BRYANT, VICTOR AND POSTILL, RONALD, EDS. *The Sunday Times Book of Brain Teasers.* New York, NY: St. Martin's Press, 1982.

*** DUDENEY, HENRY E. *536 Puzzles and Curious Problems.* New York, NY: Charles Scribner's, 1967.

** DUDENEY, HENRY E. *Amusements in Mathematics.* Mineola, NY: Dover, 1958.

DUDENEY, HENRY E. *The Canterbury Puzzles.* Mineola, NY: Dover, 1958.

DUNN, ANGELA. *Mathematical Bafflers.* Mineola, NY: Dover, 1980.

* FUJIMURA, KOBON. *The Tokyo Puzzles.* New York, NY: Charles Scribner's, 1978.

FULVES, KARL. *Self-Working Number Magic: 101 Foolproof Tricks.* Mineola, NY: Dover, 1983.

*** GARDNER, MARTIN, ED. *The Mathematical Puzzles of Sam Loyd,* 2 Vols. Mineola, NY: Dover, 1959.

HURLEY, JAMES F. *Litton's Problematical Recreations.* New York, NY: Van Nostrand Reinhold, 1971.

KORDEMSKY, BORIS A. *The Moscow Puzzles: 359 Mathematical Recreations.* New York, NY: Charles Scribner's, 1972.

LANGMAN, HARRY. *Play Mathematics.* New York, NY: Hafner Press, 1962.

** LOYD, SAM. *Sam Loyd's Cyclopedia of 5000 Puzzles, Tricks, and Conundrums.* New York, NY: Pinnacle, 1976.

MOTT-SMITH, GEOFFREY. *Mathematical Puzzles.* Mineola, NY: Dover, 1954.

* NORTHROP, EUGENE P. *Riddles in Mathematics: A Book of Paradoxes.* Melbourne, FL: Robert E. Krieger, 1975.

** PHILLIPS, HUBERT. *My Best Puzzles in Logic and Reasoning.* Mineola, NY: Dover, 1961.

** PHILLIPS, HUBERT. *My Best Puzzles in Mathematics.* Mineola, NY: Dover, 1961.

SCRIPTURE, NICHOLAS E. *Puzzles and Teasers.* London: Faber and Faber, 1970.

SHASHA, DENNIS. *The Puzzling Adventures of Dr. Ecco.* New York, NY: W.H. Freeman, 1988.

* SILVERMAN, DAVID L. *Your Move.* New York, NY: McGraw-Hill, 1971.

SMULLYAN, RAYMOND M. *Alice in Puzzleland.* New York, NY: Morrow, 1982.

SMULLYAN, RAYMOND M. *The Chess Mysteries of Sherlock Holmes.* New York, NY: Alfred A. Knopf, 1979.

* SMULLYAN, RAYMOND M. *The Lady or the Tiger? And Other Logic Puzzles.* New York, NY: Alfred A. Knopf, 1982.

* SMULLYAN, RAYMOND M. *What is the Name of This Book? The Riddle of Dracula and Other Logical Puzzles.* Englewood Cliffs, NJ: Prentice Hall, 1978.

4.4 Contests and Problems

** ALEXANDERSON, GERALD L.; KLOSINSKI, LEONARD F.; AND LARSON, LOREN C., EDS. *The William Lowell Putnam Mathematical Competition: Problems and Solutions, 1965–1984.* Washington, DC: Mathematical Association of America, 1985.

BARBEAU, EDWARD J.; KLAMKIN, MURRAY S.; AND MOSER, WILLIAM, EDS. *1001 Problems in High School Mathematics.* Montreal: Canadian Mathematical Congress, 1977.

** GLEASON, ANDREW M.; GREENWOOD, R.E.; AND KELLY, L.M. *The William Lowell Putnam Mathematical Competition: Problems and Solutions, 1938–1964.* Washington, DC: Mathematical Association of America, 1980.

* GRAHAM, L.A. *Ingenious Mathematical Problems and Methods.* Mineola, NY: Dover, 1959.

GRAHAM, L.A. *The Surprise Attack in Mathematical Problems.* Mineola, NY: Dover, 1968.

GREITZER, SAMUEL L. *International Mathematical Olympiads, 1959–1977.* Washington, DC: Mathematical Association of America, 1978.

* HARDY, KENNETH AND WILLIAMS, KENNETH S. *The Green Book: 100 Practice Problems for Undergraduate Mathematics Competitions.* Ottawa: Integer Press, 1985.

KLAMKIN, MURRAY S., ED. *Problems in Applied Mathematics: Selections from SIAM Review.* Philadelphia, PA: Society for Industrial and Applied Mathematics, 1990.

KLAMKIN, MURRAY S. *International Mathematical Olympiads, 1978–1985 and Forty Supplementary Problems.* Washington, DC: Mathematical Association of America, 1986.

KLAMKIN, MURRAY S. *U.S.A. Mathematical Olympiads, 1972–1986.* Washington, DC: Mathematical Association of America, 1988.

KÜRSCHÁK, J. *Hungarian Problem Books I and II.* Washington, DC: Mathematical Association of America, 1963.

* LARSON, LOREN C. *Problem-Solving Through Problems.* New York, NY: Springer-Verlag, 1983.

MBILI, L.S.R. *Mathematical Challenge! One Hundred Problems for the Olympiad Enthusiast.* Rondebosch, South Africa: Math Digest, 1978.

* NEWMAN, DONALD J. *A Problem Seminar.* New York, NY: Simon and Schuster, 1956–60; New York, NY: Springer-Verlag, 1982.

RUDERMAN, HARRY D., ED. *NYSML-ARML Contests, 1973–1985,* Second Edition. Reston, VA: National Council of Teachers of Mathematics, 1987.

** SALKIND, CHARLES T., *et al.*, EDS. *Contest Problem Book,* 4 Vols. Washington, DC: Mathematical Association of America, 1961–83.

YAGLOM, A.M. AND YAGLOM, I.M. *Challenging Mathematical Programs with Elementary Solutions,* 2 Vols. Mineola, NY: Dover, 1987.

4.5 Martin Gardner

GARDNER, MARTIN. *Aha! Gotcha.* New York, NY: W.H. Freeman, 1982.

GARDNER, MARTIN. *Aha! Insight.* New York, NY: W.H. Freeman, 1978.

* GARDNER, MARTIN. *Hexaflexagons and Other Mathematical Diversions.* Chicago, IL: University of Chicago Press, 1988.

* GARDNER, MARTIN. *Knotted Doughnuts and Other Mathematical Entertainments.* New York, NY: W.H. Freeman, 1986.

* GARDNER, MARTIN. *Martin Gardner's New Mathematical Diversions from Scientific American.* Chicago, IL: University of Chicago Press, 1983.

* GARDNER, MARTIN. *Martin Gardner's Sixth Book of Mathematical Diversions from Scientific American.* Chicago, IL: University of Chicago Press, 1983.

* GARDNER, MARTIN. *Mathematical Carnival.* Washington, DC: Mathematical Association of America, 1989.

* GARDNER, MARTIN. *Mathematical Circus: More Games, Puzzles, Paradoxes, and Other Mathematical Entertainments from Scientific American.* New York, NY: Alfred A. Knopf, 1979.

* GARDNER, MARTIN. *Mathematical Magic Show.* New York, NY: Alfred A. Knopf, 1977; Washington, DC: Mathematical Association of America, 1990.

GARDNER, MARTIN. *Mathematics, Magic and Mystery.* Mineola, NY: Dover, 1956.

* GARDNER, MARTIN. *Penrose Tiles to Trapdoor Ciphers.* New York, NY: W.H. Freeman, 1989.

* GARDNER, MARTIN. *Riddles of the Sphinx And Other Mathematical Puzzle Tales.* Washington, DC: Mathematical Association of America, 1987.

GARDNER, MARTIN. *Science Fiction Puzzle Tales.* New York, NY: Clarkson N. Potter, 1981.

* GARDNER, MARTIN. *The Magic Numbers of Dr. Matrix.* Buffalo, NY: Prometheus Books, 1985.

* GARDNER, MARTIN. *The Scientific American Book of Mathematical Puzzles and Diversions.* New York, NY: Simon and Schuster, 1959.

* GARDNER, MARTIN. *The Second Scientific American Book of Mathematical Puzzles and Diversions.* New York, NY: Simon and Schuster, 1961.

* GARDNER, MARTIN. *Time Travel and Other Mathematical Bewilderments.* New York, NY: W.H. Freeman, 1987.

* GARDNER, MARTIN. *Wheels, Life, and Other Mathematical Amusements.* New York, NY: W.H. Freeman, 1983.

4.6 Miscellany

* GARDINER, A. *Mathematical Puzzling.* New York, NY: Oxford University Press, 1987.

** GOLOMB, SOLOMON W. *Polyominoes.* New York, NY: Charles Scribner's, 1965.

GREENBLATT, M.H. *Mathematical Entertainments.* London: George Allen and Unwin, 1968.

HEAFFORD, PHILIP. *The Math Entertainer.* New York, NY: Vintage Press, 1983.

KLARNER, DAVID A., ED. *The Mathematical Gardner.* Boston, MA: Prindle, Weber and Schmidt, 1981.

** LINDGREN, HARRY. *Geometric Dissections.* New York, NY: Van Nostrand Reinhold, 1964.

MADACHY, JOSEPH S. *Mathematics on Vacation.* New York, NY: Charles Scribner's, 1966.

* MAXWELL, E.A. *Fallacies in Mathematics.* New York, NY: Cambridge University Press, 1959.

SCHATTSCHNEIDER, DORIS J. AND WALKER, WALLACE. *M.C. Escher Kaleidocycles.* New York, NY: Ballantine Books, 1978.

SIMON, WILLIAM. *Mathematical Magic.* New York, NY: Charles Scribner's, 1964.

* SMULLYAN, RAYMOND M. *Forever Undecided: A Puzzle Guide to Gödel.* New York, NY: Alfred A. Knopf, 1987.

* SMULLYAN, RAYMOND M. *To Mock a Mockingbird.* New York, NY: Alfred A. Knopf, 1985.

5 Education

5.1 Policy

* COCKCROFT, WILFRED H., ED. *Mathematics Counts.* London: Her Majesty's Stationery Office, 1982.

ELLERTON, N.F. AND CLEMENTS, M.A., EDS. *School Mathematics: The Challenge to Change.* Geelong, Australia: Deakin University, 1989.

* HOWSON, A. GEOFFREY AND WILSON, BRYAN. *School Mathematics in the 1990s.* New York, NY: Cambridge University Press, 1986.

* HOWSON, A. GEOFFREY AND KAHANE, J.-P., EDS. *The Popularization of Mathematics.* New York, NY: Cambridge University Press, 1990.

* KILPATRICK, JEREMY. *Academic Preparation in Mathematics: Teaching for Transition From High School to College.* New York, NY: College Board, 1985.

MADISON, BERNARD L. AND HART, THERESE A. *A Challenge of Numbers: People in the Mathematical Sciences.* Washington, DC: National Academy Press, 1990.

** MATHEMATICAL SCIENCES EDUCATION BOARD. *Reshaping School Mathematics: A Philosophy and Framework for Curriculum.* Washington, DC: National Academy Press, 1990.

MESTRE, JOSE P. AND LOCHHEAD, JACK. *Academic Preparation In Science: Teaching for Transition From High School To College,* Second Edition. New York, NY: College Board, 1990.

*** NATIONAL RESEARCH COUNCIL. *Everybody Counts: A Report to the Nation on the Future of Mathematics Education.* Washington, DC: National Academy Press, 1989.

** NATIONAL RESEARCH COUNCIL. *Moving Beyond Myths: Revitalizing Undergraduate Mathematics.* Washington, DC: National Academy Press, 1991.

* NATIONAL RESEARCH COUNCIL. *Renewing U.S. Mathematics: A Plan for the 1990s.* Washington, DC: National Academy Press, 1990.

* NATIONAL RESEARCH COUNCIL. *Renewing U.S. Mathematics: Critical Resource for the Future.* Washington, DC: National Research Council, 1984.

PRICE, JACK AND GAWRONSKI, J.D., EDS. *Changing School Mathematics: A Responsive Process.* Reston, VA: National Council of Teachers of Mathematics, 1981.

ROMBERG, THOMAS A. *School Mathematics: Options for the 1990's.* Washington, DC: U.S. Government Printing Office, 1984.

*** STEEN, LYNN ARTHUR, ED. *On the Shoulders of Giants: New Approaches to Numeracy.* Washington, DC: National Academy Press, 1990.

5.2 Philosophy

* CHRISTIANSEN, B., *et al.* *Perspectives on Mathematics Education.* Norwell, MA: D. Reidel, 1986.

** DE LANGE, JAN. *Mathematics, Insight, and Meaning.* Utrecht: Rijksuniversiteit Utrecht, 1987.

** ERNEST, PAUL. *The Philosophy of Mathematics Education.* New York, NY: Falmer Press, 1991.

FREUDENTHAL, HANS. *Didactical Phenomenology of Mathematical Structures.* Norwell, MA: D. Reidel, 1983.

FREUDENTHAL, HANS. *Mathematics as an Educational Task.* Norwell, MA: D. Reidel, 1973.

** HIEBERT, JAMES, ED. *Conceptual and Procedural Knowledge: The Case of Mathematics.* Hillsdale, NJ: Lawrence Erlbaum, 1986.

* MELLIN-OLSEN, STEIG. *The Politics of Mathematics Education: Mathematics Education Library.* Norwell, MA: D. Reidel, 1987.

5.3 Psychology

FISCHBEIN, E. *The Intuitive Sources of Probabilistic Thinking in Children.* Norwell, MA: D. Reidel, 1975.

FUYS, DAVID; GEDDES, DOROTHY; AND TISCHLER, ROSAMOND. *The Van Hiele Model of Thinking in Geometry Among Adolescents.* Reston, VA: National Council of Teachers of Mathematics, 1988.

GINSBURG, H.P., ED. *The Development of Mathematical Thinking.* New York, NY: Academic Press, 1983.

*** HADAMARD, JACQUES. *Psychology of Invention in the Mathematical Field.* Mineola, NY: Dover, 1945.

INHELDER, B. AND PIAGET, JEAN. *The Growth of Logical Thinking From Childhood to Adolescence.* New York, NY: Basic Books, 1958.

JANVIER, CLAUDE, ED. *Problems of Representation in the Teaching and Learning of Mathematics.* Hillsdale, NJ: Lawrence Erlbaum, 1987.

* LAVE, J. *Cognition in Practice: Mind, Mathematics, and Culture in Everyday Life.* New York, NY: Cambridge University Press, 1988.

LESH, R. AND LANDAU, M., EDS. *Acquisition of Mathematics Concepts and Processes.* New York, NY: Academic Press, 1983.

NESHER, PEARLA AND KILPATRICK, JEREMY, EDS. *Mathematics and Cognition.* New York, NY: Cambridge University Press, 1990.

PIAGET, JEAN.; INHELDER, B.; AND SZEMINSKA, A. *The Child's Conception of Geometry.* New York, NY: Harper Torchbooks, 1964.

* PIAGET, JEAN. *The Child's Conception of Number.* New York, NY: W.W. Norton, 1965.

** RESNICK, LAUREN B. AND FORD, WENDY W. *The Psychology of Mathematics for Instruction.* Hillsdale, NJ: Lawrence Erlbaum, 1981.

* SCHOENFELD, ALAN H., ED. *Cognitive Science and Mathematics Education.* Hillsdale, NJ: Lawrence Erlbaum, 1987.

** SKEMP, RICHARD R. *The Psychology of Learning Mathematics,* Second Edition. New York, NY: Penguin, 1973; Hillsdale, NJ: Lawrence Erlbaum, 1986.

VERGNAUD, G. *L'enfant, la mathématique et la réalité.* Bern: Peter Lang, 1981.

5.4 Culture

ASCHER, MARCIA AND ASCHER, ROBERT. *Code of the Quipu: A Study in Media, Mathematics, and Culture.* Ann Arbor, MI: University of Michigan Press, 1981.

* ASCHER, MARCIA. *Ethnomathematics: A Multicultural View of Mathematical Ideas.* Pacific Grove, CA: Brooks/Cole, 1991.

* BISHOP, ALAN J. *Mathematical Enculturation: A Cultural Perspective on Mathematics Education.* Norwell, MA: Kluwer Academic, 1988.

CLOSS, MICHAEL P., ED. *Native American Mathematics.* Austin, TX: University of Texas Press, 1986.

COCKING, R.R. AND MESTRE, JOSE P., EDS. *Linguistic and Cultural Influences on Learning Mathematics.* Hillsdale, NJ: Lawrence Erlbaum, 1988.

ERNEST, PAUL, ED. *The Social Context of Mathematics Teaching.* Exeter: University of Exeter Press, 1988.

* ZASLAVSKY, CLAUDIA. *Africa Counts: Number and Pattern in African Culture.* Boston, MA: Prindle, Weber and Schmidt, 1973.

5.5 History

** BAUMGART, JOHN K. *Historical Topics for the Mathematics Classroom.* Reston, VA: National Council of Teachers of Mathematics, 1969, 1989.

* BIDWELL, JAMES K. AND CLASON, ROBERT G., EDS. *Readings in the History of Mathematics Education.* Reston, VA: National Council of Teachers of Mathematics, 1970.

CAJORI, FLORIAN. *The Teaching and History of Mathematics in the United States.* Wilmington, DE: Scholarly Resources, 1974.

* COHEN, PATRICIA CLINE. *A Calculating People: The Spread of Numeracy in Early America.* Chicago, IL: University of Chicago Press, 1982.

GRATTAN-GUINNESS, I., ED. *History in Mathematics Education.* Paris: Belin, 1986.

JONES, PHILLIP S., ED. *A History of Mathematics Education in the United States and Canada: 32nd Yearbook.* Reston, VA: National Council of Teachers of Mathematics, 1970.

MOON, B. *The "New Maths" Curriculum Controversy: An International Story.* New York, NY: Falmer Press, 1986.

5.6 Curriculum and Instruction

* AMERICAN ASSOCIATION FOR THE ADVANCEMENT OF SCIENCE. *Science For All Americans.* Washington, DC: American Association for the Advancement of Science, 1989.

ATHEN, HERMANN AND HEINZ, KUNLE, EDS. *Proceedings of the Third International Congress on Mathematical Education.* Karlsruhe, Germany: Zentralblatt für Didaktik der Mathematik, 1977.

* BLACKWELL, DAVID AND HENKIN, LEON. *Mathematics: Report of the Project 2061 Phase I Mathematics Panel.* Washington, DC: American Association for the Advancement of Science, 1989.

CARSS, MARJORIE, ED. *Proceedings of the Fifth International Congress on Mathematical Education.* New York, NY: Birkhäuser, 1986.

* COMMITTEE ON MATHEMATICAL EDUCATION OF TEACHERS. *Guidelines for the Continuing Mathematical Education of Teachers.* Washington, DC: Mathematical Association of America, 1988.

CONFERENCE BOARD OF THE MATHEMATICAL SCIENCES. *Overview and Analysis of School Mathematics, Grades K–12.* Washington, DC: Conference Board of the Mathematical Sciences, 1975.

*** COONEY, THOMAS J., ED. *Teaching and Learning Mathematics in the 1990s: 1990 Yearbook.* Reston, VA: National Council of Teachers of Mathematics, 1990.

* DAVIDSON, NEIL, ED. *Cooperative Learning in Mathematics: A Handbook for Teachers.* Reading, MA: Addison-Wesley, 1990.

DRISCOLL, MARK J. AND CONFREY, J., EDS. *Teaching Mathematics: Strategies that Work.* Chelmsford, MA: Northeast Regional Exchange, 1985.

ERNEST, PAUL. *Mathematics Teaching: The State of the Art.* New York, NY: Falmer Press, 1989.

HIRST, ANN AND HIRST, KEITH, EDS. *Proceedings of the Sixth International Congress on Mathematical Education.* Budapest: János Bolyai Mathematical Society, 1988.

HOWSON, A. GEOFFREY, ED. *Developments in Mathematical Education: Proceedings of the Second International Congress on Mathematical Education.* New York, NY: Cambridge University Press, 1973.

* HOWSON, A. GEOFFREY, *et al.*, EDS. *Mathematics as a Service Subject.* New York, NY: Cambridge University Press, 1988.

HOWSON, A. GEOFFREY; KEITEL, CHRISTINE; AND KILPATRICK, JEREMY. *Curriculum Development in Mathematics.* New York, NY: Cambridge University Press, 1981.

HOWSON, A. GEOFFREY. *Challenges and Responses in Mathematics.* New York, NY: Cambridge University Press, 1987.

* HOWSON, A. GEOFFREY. *National Curricula in Mathematics.* Leicester, England: The Mathematical Association, 1991.

** LINDQUIST, MARY M. AND SHULTE, ALBERT P. *Learning and Teaching Geometry, K–12: 1987 Yearbook.* Reston, VA: National Council of Teachers of Mathematics, 1987.

NATIONAL COUNCIL OF TEACHERS OF MATHEMATICS. *A Sourcebook of Applications of School Mathematics.* Reston, VA: National Council of Teachers of Mathematics, 1980.

*** NATIONAL COUNCIL OF TEACHERS OF MATHEMATICS. *Curriculum and Evaluation Standards for School Mathematics.* Reston, VA: National Council of Teachers of Mathematics, 1989.

*** NATIONAL COUNCIL OF TEACHERS OF MATHEMATICS. *Professional Standards for Teaching Mathematics.* Reston, VA: National Council of Teachers of Mathematics, 1991.

* SCHOEN, HAROLD L. AND ZWENG, MARILYN J. *Estimation and Mental Computation: 1986 Yearbook.* Reston, VA: National Council of Teachers of Mathematics, 1986.

* SHULTE, ALBERT P., ED. *Teaching Statistics and Probability: 1981 Yearbook.* Reston, VA: National Council of Teachers of Mathematics, 1981.

* SILVER, EDWARD A.; KILPATRICK, JEREMY; AND SCHLESINGER, B. *Thinking Through Mathematics: Fostering Inquiry and Communication in Mathematics Classrooms.* New York, NY: College Board, 1990.

STEEN, LYNN ARTHUR AND ALBERS, DONALD J., EDS. *Teaching Teachers, Teaching Students: Reflections on Mathematical Education.* New York, NY: Birkhäuser, 1981.

SWAN, M., ED. *The Language of Functions and Graphs.* Nottingham: Shell Centre for Mathematical Education, 1985.

* TAYLOR, ROSS, ED. *Professional Development for Teachers of Mathematics: A Handbook.* Reston, VA: National Council of Teachers of Mathematics, 1986.

TOBIAS, SHEILA. *Overcoming Math Anxiety.* New York, NY: W.W. Norton, 1978; Boston, MA: Houghton Mifflin, 1980.

TOBIAS, SHEILA. *Succeed With Math: Every Student's Guide to Conquering Math Anxiety.* New York, NY: College Board, 1987.

ZWENG, MARILYN J., *et al.*, EDS. *Proceedings of the Fourth International Congress on Mathematical Education.* New York, NY: Birkhäuser, 1983.

5.7 Elementary Education

BAROODY, A.J. *A Guide to Teaching Mathematics in the Primary Grades.* Needham Heights, MA: Allyn and Bacon, 1989.

BURNS, MARILYN. *The Book of Think (Or How to Solve a Problem Twice Your Size).* Waltham, MA: Little, Brown, 1976.

BURNS, MARILYN. *The I Hate Mathematics! Book.* Waltham, MA: Little, Brown, 1975.

FENNELL, FRANCIS AND WILLIAMS, DAVID E. *Ideas from the Arithmetic Teacher, Grades 4–6: Intermediate School,* Second Edition. Reston, VA: National Council of Teachers of Mathematics, 1986.

GINSBURG, H.P. *Children's Arithmetic: The Learning Process.* New York, NY: Van Nostrand Reinhold, 1977.

HILL, JANE M., ED. *Geometry for Grades K–6: Readings from the Arithmetic Teacher.* Reston, VA: National Council of Teachers of Mathematics, 1987.

IMMERZEEL, GEORGE AND THOMAS, MELVIN. *Ideas from the Arithmetic Teacher, Grades 1–4: Primary.* Reston, VA: National Council of Teachers of Mathematics, 1982.

** PAYNE, JOSEPH N., ED. *Mathematics For the Young Child.* Reston, VA: National Council of Teachers of Mathematics, 1990.

POST, THOMAS R., ED. *Teaching Mathematics in Grades K–8: Research Based Methods.* Needham Heights, MA: Allyn and Bacon, 1988.

REYS, ROBERT E.; SUYDAM, MARILYN N.; AND LINDQUIST, MARY M. *Helping Children Learn Mathematics.* Englewood Cliffs, NJ: Prentice Hall, 1984.

SILVEY, LINDA AND SMART, JAMES R. *Mathematics for the Middle Grades (5–9): 1982 Yearbook.* Reston, VA: National Council of Teachers of Mathematics, 1982.

* STEFFE, LESLIE P. AND WOOD, TERRY, EDS. *International Perspectives on Transforming Early Childhood Mathematics Education.* Hillsdale, NJ: Lawrence Erlbaum, 1990.

* STENMARK, JEAN K.; THOMPSON, VIRGINIA; AND COSSEY, RUTH. *Family Math.* Berkeley, CA: Lawrence Hall of Science, 1986.

*** TRAFTON, PAUL R. AND SHULTE, ALBERT P., EDS. *New Directions for Elementary School Mathematics: 1989 Yearbook.* Reston, VA: National Council of Teachers of Mathematics, 1989.

* VAN DE WALLE, JOHN A. *Elementary School Mathematics: Teaching Developmentally.* White Plains, NY: Longman, 1990.

WEISSGLASS, JULIAN. *Exploring Elementary Mathematics: A Small-Group Approach for Teaching,* Second Edition. New York, NY: W.H. Freeman, 1979; Dubuque, IA: Kendall-Hunt, 1990.

* WORTH, JOAN, ED. *Preparing Elementary School Mathematics Teachers: Readings from the Arithmetic Teacher.* Reston, VA: National Council of Teachers of Mathematics, 1987.

5.8 Secondary Education

BIGGS, E. *Teaching Mathematics 7–13: Slow Learning and Able Pupils.* Windsor, England: NFER-Nelson, 1985.

* CAMPBELL, PAUL J. AND GRINSTEIN, LOUISE S. *Mathematics Education in Secondary Schools and Two-Year Colleges: A Sourcebook.* New York, NY: Garland, 1988.

* COOPER, B. *Renegotiating Secondary School Mathematics: A Study of Curriculum Change and Stability.* New York, NY: Falmer Press, 1985.

*** COXFORD, ARTHUR F. AND SHULTE, ALBERT P. *The Ideas of Algebra, K–12: 1988 Yearbook.* Reston, VA: National Council of Teachers of Mathematics, 1988.

DALTON, LEROY C. AND SNYDER, HENRY D., EDS. *Topics for Mathematics Clubs.* Reston, VA: National Council of Teachers of Mathematics, 1983.

EASTERDAY, KENNETH E.; HENRY, LOREN L.; AND SIMPSON, F. MORGAN. *Activities for Junior High School and Middle School Mathematics.* Reston, VA: National Council of Teachers of Mathematics, 1981.

** HIRSCH, CHRISTIAN R. AND ZWENG, MARILYN J. *The Secondary School Mathematics Curriculum: 1985 Yearbook.* Reston, VA: National Council of Teachers of Mathematics, 1985.

HOLMES, PETER, ED. *The Best of "Teaching Statistics."* Sheffield, England: Teaching Statistics Trust, 1986.

* IMMERZEEL, GEORGE AND THOMAS, MELVIN. *Ideas from the Arithmetic Teacher, Grades 6–8: Middle School.* Reston, VA: National Council of Teachers of Mathematics, 1982.

SOBEL, MAX A., ED. *Readings for Enrichment in Secondary School Mathematics.* Reston, VA: National Council of Teachers of Mathematics, 1988.

5.9 Undergraduate Education

** ALBERS, DONALD J., *et al.* *A Statistical Abstract of Undergraduate Programs in the Mathematical and Computer Sciences, 1990–91.* Washington, DC: Mathematical Association of America, 1992.

ALBERS, DONALD J.; RODI, STEPHEN B.; AND WATKINS, ANN E., EDS. *New Directions in Two-Year College Mathematics.* New York, NY: Springer-Verlag, 1985.

COMMITTEE ON THE UNDERGRADUATE PROGRAM IN MATHEMATICS. *A Compendium of CUPM Recommendations,* 2 Vols. Washington, DC: Mathematical Association of America, 1975.

COMMITTEE ON THE UNDERGRADUATE PROGRAM IN MATHEMATICS. *Recommendations for a General Mathematical Sciences Program.* Washington, DC: Mathematical Association of America, 1981.

** COMMITTEE ON THE UNDERGRADUATE PROGRAM IN MATHEMATICS. *Reshaping College Mathematics.* Washington, DC: Mathematical Association of America, 1989.

DAVIS, RONALD M., ED. *A Curriculum in Flux: Mathematics at Two-Year Colleges.* Washington, DC: Mathematical Association of America, 1989.

** DOUGLAS, RONALD G., ED. *Toward A Lean and Lively Calculus.* Washington, DC: Mathematical Association of America, 1986.

KLINE, MORRIS. *Why the Professor Can't Teach: Mathematics and the Dilemma of University Education.* New York, NY: St. Martin's Press, 1977.

* KNUTH, DONALD E.; LARRABEE, TRACY; AND ROBERTS, PAUL M. *Mathematical Writing.* Washington, DC: Mathematical Association of America, 1989.

* LEINBACH, L. CARL, *et al.* *The Laboratory Approach to Teaching Calculus.* Washington, DC: Mathematical Association of America, 1991.

*** LEITZEL, JAMES R.C., ED. *A Call For Change: Recommendations for the Mathematical Preparation of Teachers of Mathematics.* Washington, DC: Mathematical Association of America, 1991.

RALSTON, ANTHONY AND YOUNG, GAIL S., EDS. *The Future of College Mathematics.* New York, NY: Springer-Verlag, 1983.

RALSTON, ANTHONY, ED. *Discrete Mathematics in the First Two Years.* Washington, DC: Mathematical Association of America, 1989.

* SCHOENFELD, ALAN H., ED. *A Sourcebook for College Mathematics Teaching.* Washington, DC: Mathematical Association of America, 1990.

SENECHAL, LESTER, ED. *Models for Undergraduate Research in Mathematics.* Washington, DC: Mathematical Association of America, 1990.

SMITH, DAVID A., *et al.*, EDS. *Computers and Mathematics: The Use of Computers in Undergraduate Instruction.* Washington, DC: Mathematical Association of America, 1988.

** STEEN, LYNN ARTHUR, ED. *Calculus for a New Century: A Pump, Not a Filter.* Washington, DC: Mathematical Association of America, 1988.

* STERRETT, ANDREW, ED. *Using Writing to Teach Mathematics.* Washington, DC: Mathematical Association of America, 1990.

** TOBIAS, SHEILA. *They're Not Dumb, They're Different: Stalking the Second Tier.* Tucson, AZ: Research Corporation, 1990.

** TUCKER, THOMAS W., ED. *Priming the Calculus Pump: Innovations and Resources.* Washington, DC: Mathematical Association of America, 1990.

* ZIMMERMANN, WALTER AND CUNNINGHAM, STEVE, EDS. *Visualization in Teaching and Learning Mathematics.* Washington, DC: Mathematical Association of America, 1991.

5.10 Minorities and Women

BURTON, L., ED. *Girls Into Maths Can Go.* Bristol, PA: Taylor and Francis, 1986.

CROWLEY, MICHAEL F. AND LANE, MELISSA J., EDS. *Women and Minorities in Science and Engineering.* Washington, DC: National Science Foundation, 1986.

** FENNEMA, ELIZABETH AND LEDER, GILAH, EDS. *Mathematics and Gender.* New York, NY: Teachers College Press, 1990.

** GRINSTEIN, LOUISE S. AND CAMPBELL, PAUL J., EDS. *Women of Mathematics: A Biobibliographic Sourcebook.* Westport, CT: Greenwood Press, 1987.

HØYRUP, ELSE. *Women and Mathematics, Science, and Engineering.* Roskilde: Roskilde University Center, 1978.

* KENSCHAFT, PATRICIA C. AND KEITH, SANDRA, EDS. *Winning Women Into Mathematics.* Washington, DC: Mathematical Association of America, 1991.

* MOZANS, H.J. *Woman in Science.* New York, NY: Appleton, 1913; Cambridge, MA: MIT Press, 1974; Notre Dame, IN: University of Notre Dame Press, 1991.

NEWELL, VIRGINIA K., *et al.*, EDS. *Black Mathematicians and Their Works.* Ardmore, PA: Dorrance, 1980.

* ORR, ELEANOR WILSON. *Twice As Less: Black English and the Performance of Black Students in Mathematics and Science.* New York, NY: W.W. Norton, 1987.

OSEN, L.M. *Women in Mathematics.* Cambridge, MA: MIT Press, 1974.

PERL, T.H. *Math Equals.* Reading, MA: Addison-Wesley, 1978.

SAMMONS, VIVIAN O. *Blacks in Science and Medicine.* New York, NY: Hemisphere, 1990.

5.11 Problem Solving

BROWN, STEPHEN I. AND WALTER, MARION I. *The Art of Problem Posing.* Philadelphia, PA: Franklin Institute Press, 1983.

CHARLES, RANDALL I.; LESTER, FRANK; AND O'DAFFER, PHARES G. *How To Evaluate Progress in Problem Solving.* Reston, VA: National Council of Teachers of Mathematics, 1987.

* MASON, JOHN H.; BURTON, LEONE; AND STACEY, KAYE. *Thinking Mathematically.* Reading, MA: Addison-Wesley, 1982, 1985.

* MASON, JOHN H. *Learning and Doing Mathematics.* Houndmills, England: Macmillan Education, 1988.

MCLEOD, DOUGLAS AND ADAMS, V., EDS. *Affect and Mathematical Problem Solving: A New Perspective.* New York, NY: Springer-Verlag, 1989.

O'DAFFER, PHARES G., ED. *Problem Solving Tips for Teachers: Selections from the Arithmetic Teacher.* Reston, VA: National Council of Teachers of Mathematics, 1988.

*** PÓLYA, GEORGE. *How To Solve It,* Second Edition. Princeton, NJ: Princeton University Press, 1945; New York, NY: Doubleday, 1957.

*** PÓLYA, GEORGE. *Mathematical Discovery: On Understanding, Learning, and Teaching Problem Solving,* Combined Edition. New York, NY: John Wiley, 1962, 1981.

** PÓLYA, GEORGE. *Mathematics and Plausible Reasoning,* 2 Vols. Princeton, NJ: Princeton University Press, 1954, 1969.

RUBINSTEIN, MOSHE F. *Patterns of Problem Solving.* Englewood Cliffs, NJ: Prentice Hall, 1975.

RUBINSTEIN, MOSHE F. *Tools for Thinking and Problem Solving.* Englewood Cliffs, NJ: Prentice Hall, 1986.

*** SCHOENFELD, ALAN H. *Mathematical Problem Solving.* New York, NY: Academic Press, 1985.

* SCHOENFELD, ALAN H. *Problem Solving in the Mathematics Curriculum: A Report, Recommendations, and an Annotated Bibliography.* Washington, DC: Mathematical Association of America, 1983.

SILVER, EDWARD A., ED. *Teaching and Learning Mathematical Problem Solving: Multiple Research Perspectives.* Hillsdale, NJ: Lawrence Erlbaum, 1985.

5.12 Research and Research Summaries

* AMERICAN EDUCATION RESEARCH ASSOCIATION. *Research in Teaching and Learning,* Vol. 4: Mathematics and Natural Sciences. New York, NY: Macmillan, 1986.

BELL, A.W.; COSTELLO, J.; AND KUCHEMANN, D. *A Review of Research in Mathematical Education: Research on Learning and Teaching.* Windsor, England: NFER-Nelson, 1983.

CARPENTER, THOMAS; MOSER, JAMES; AND ROMBERG, THOMAS A., EDS. *Addition and Subtraction: A Cognitive Perspective.* Hillsdale, NJ: Lawrence Erlbaum, 1982.

** CHARLES, RANDALL I. AND SILVER, EDWARD A., EDS. *The Teaching and Assessing of Mathematical Problem Solving.* Hillsdale, NJ: Lawrence Erlbaum, 1988; Reston, VA: National Council of Teachers of Mathematics, 1988.

DRISCOLL, MARK J. *Research Within Reach: Elementary School Mathematics.* Reston, VA: National Council of Teachers of Mathematics, 1982.

* FUSON, KAREN C. *Children's Counting and Concepts of Number.* New York, NY: Springer-Verlag, 1988.

** GROUWS, DOUGLAS A.; COONEY, THOMAS J.; AND JONES, DOUGLAS, EDS. *Perspectives on Research on Effective Mathematics Teaching.* Hillsdale, NJ: Lawrence Erlbaum, 1988; Reston, VA: National Council of Teachers of Mathematics, 1988.

** HIEBERT, JAMES AND BEHR, MERLYN, EDS. *Number Concepts and Operations in the Middle Grades.* Hillsdale, NJ: Lawrence Erlbaum, 1988; Reston, VA: National Council of Teachers of Mathematics, 1988.

MURNANE, RICHARD J. AND RAIZEN, SENTA A., EDS. *Improving Indicators of the Quality of Science and Mathematics Education in Grades K–12.* Washington, DC: National Academy Press, 1988.

* RESNICK, LAUREN B. AND KLOPFER, L.E., EDS. *Toward the Thinking Curriculum: Current Cognitive Research.* Alexandria, VA: Association for Supervision and Curriculum Development, 1989.

*** RESNICK, LAUREN B. *Education and Learning to Think.* Washington, DC: National Academy Press, 1987.

SHUMWAY, RICHARD J., ED. *Research in Mathematics Education.* Reston, VA: National Council of Teachers of Mathematics, 1980.

** SOWDER, JUDITH T., ED. *Setting a Research Agenda.* Hillsdale, NJ: Lawrence Erlbaum, 1989; Reston, VA: National Council of Teachers of Mathematics, 1989.

STEFFE, LESLIE P.; COBB, PAUL; AND VON GLASERFELD, ERNST. *Construction of Arithmetical Meanings and Strategies.* New York, NY: Springer-Verlag, 1988.

SUYDAM, MARILYN N. AND DESSART, DONALD J. *Classroom Ideas from Research on Secondary School Mathematics.* Reston, VA: National Council of Teachers of Mathematics, 1983.

** WAGNER, SIGRID AND KIERAN, CAROLYN, EDS. *Research Issues in the Learning and Teaching of Algebra.* Hillsdale, NJ: Lawrence Erlbaum, 1989; Reston, VA: National Council of Teachers of Mathematics, 1989.

5.13 Studies and Assessment

CALIFORNIA ASSESSMENT PROGRAM. *A Question of Thinking: A First Look at Students' Performance on Open-ended Questions in Mathematics.* Sacramento, CA: California State Department of Education, 1989.

** DOSSEY, JOHN A., *et al.* *The Mathematics Report Card: Are We Measuring Up?* Princeton, NJ: Educational Testing Service, 1988.

KULM, GERALD, ED. *Assessing Higher Order Thinking in Mathematics.* Washington, DC: American Association for the Advancement of Science, 1990.

* LAPOINTE, ARCHIE; MEAD, NANCY; AND PHILLIPS, GARY. *A World of Differences: An International Assessment of Mathematics and Science.* Princeton, NJ: Educational Testing Service, 1989.

* LINDQUIST, MARY M., ED. *Results from the Fourth Mathematics Assessment of the National Assessment of Educational Progress.* Reston, VA: National Council of Teachers of Mathematics, 1989.

** MCKNIGHT, CURTIS C., *et al.* *The Underachieving Curriculum: Assessing U.S. School Mathematics From An International Perspective.* Champaign, IL: Stipes, 1987.

ROBITAILLE, DAVID F. AND GARDEN, ROBERT A. *The IEA Study of Mathematics II: Contexts and Outcomes of School Mathematics.* Elmsford, NY: Pergamon Press, 1989.

* ROMBERG, THOMAS A., ED. *Mathematics Assessment and Evaluation: Imperatives for Mathematics Educators.* Albany, NY: SUNY Press, 1991.

* STEVENSON, HAROLD W., *et al.* *Making the Grade in Mathematics: Elementary School Mathematics in the United States, Taiwan, and Japan.* Reston, VA: National Council of Teachers of Mathematics, 1990.

* STIGLER, JAMES W.; LEE, SHIN-YING; AND STEVENSON, HAROLD W. *Mathematical Knowledge of Japanese, Chinese, and American Elementary School Children.* Reston, VA: National Council of Teachers of Mathematics, 1990.

TRAVERS, KENNETH J. AND WESTBURY, IAN, EDS. *The IEA Study of Mathematics I: Analysis of Mathematics Curricula.* Elmsford, NY: Pergamon Press, 1989.

TRAVERS, KENNETH J., ED. *Second International Mathematics Study: Detailed Report for the United States.* Champaign, IL: Stipes, 1986.

5.14 Computers and Technology

* CHAZAN, DANIEL AND HOUDE, RICHARD. *How to Use Conjecturing and Microcomputers to Teach Geometry.* Reston, VA: National Council of Teachers of Mathematics, 1989.

FEY, JAMES T., ED. *Computing and Mathematics: The Impact on Secondary School Curricula.* Reston, VA: National Council of Teachers of Mathematics, 1984.

HANSEN, VIGGO P. AND ZWENG, MARILYN J. *Computers in Mathematics Education: 1984 Yearbook.* Reston, VA: National Council of Teachers of Mathematics, 1984.

HOWSON, A. GEOFFREY AND KAHANE, J.-P., EDS. *The Influence of Computers and Informatics on Mathematics and Its Teaching.* New York, NY: Cambridge University Press, 1986.

** PAPERT, SEYMOUR. *Mindstorms: Children, Computers, and Powerful Ideas.* New York, NY: Basic Books, 1980.

6 Calculus and Precalculus

6.1 School Mathematics

ALLENDOERFER, C.B. AND OAKLEY, C.O. *Principles of Mathematics.* New York, NY: McGraw-Hill, 1963.

AUSLANDER, LOUIS. *What Are Numbers?* Glenview, IL: Scott Foresman, 1969.

* MARTIN, EDWARD, ED. *Elements of Mathematics, Book B: Problem Book.* St. Louis, MO: CEMREL-CSMP, 1975.

* RISING, GERALD R., ED. *Unified Mathematics,* 3 Vols. Boston, MA: Houghton Mifflin, 1981.

SEYMOUR, DALE. *Visual Patterns in Pascal's Triangle.* Palo Alto, CA: Dale Seymour, 1986.

WEBBER, G. AND BROWN, J. *Basic Concepts of Mathematics.* Reading, MA: Addison-Wesley, 1963.

6.2 Precalculus

ALLENDOERFER, C.B. AND OAKLEY, C.O. *Fundamentals of Freshman Mathematics,* Second Edition. New York, NY: McGraw-Hill, 1965.

AYRE, H.G.; STEPHENS, R.; AND MOCK, G.D. *Analytic Geometry: Two and Three Dimensions,* Second Edition. New York, NY: Van Nostrand Reinhold, 1967.

COHEN, DAVID. *Precalculus,* Third Edition. St. Paul, MN: West, 1984, 1989.

COXFORD, ARTHUR F. AND PAYNE, JOSEPH N. *Advanced Mathematics: A Preparation for Calculus.* San Diego, CA: Harcourt Brace Jovanovich, 1984.

DEMANA, FRANKLIN D. AND LEITZEL, JOAN R. *Transition to College Mathematics.* Reading, MA: Addison-Wesley, 1984.

*** DEMANA, FRANKLIN D. AND WAITS, BERT K. *Precalculus Mathematics—A Graphing Approach.* Reading, MA: Addison-Wesley, 1990.

*** DEMANA, FRANKLIN D., *et al.* *Graphing Calculator and Computer Graphing Laboratory Manual,* Second Edition. Reading, MA: Addison-Wesley, 1991.

DEVLIN, KEITH J. *Sets, Functions, and Logic: Basic Concepts of University Mathematics.* New York, NY: Chapman and Hall, 1981.

* FOERSTER, PAUL A. *Precalculus with Trigonometry: Functions and Applications.* Reading, MA: Addison-Wesley, 1987.

GROSSMAN, STANLEY I. *Algebra and Trigonometry.* Philadelphia, PA: Saunders College, 1989.

* KAUFMANN, JEROME E. *College Algebra and Trigonometry,* Second Edition. Boston, MA: PWS-Kent, 1987, 1990.

LARSON, LOREN C. *Algebra and Trigonometry Refresher for Calculus Students.* New York, NY: W.H. Freeman, 1979.

** LEITHOLD, LOUIS. *Before Calculus: Functions, Graphs, and Analytic Geometry,* Second Edition. New York, NY: Harper and Row, 1985, 1989.

LEWIS, PHILIP G. *Approaching Precalculus Mathematics Discretely: Explorations in a Computer Environment.* Cambridge, MA: MIT Press, 1990.

SIMMONS, GEORGE F. *Precalculus Mathematics in a Nutshell: Geometry, Algebra, Trigonometry.* Los Altos, CA: William Kaufmann, 1981.

* SOBEL, MAX A. AND LERNER, NORBERT. *Algebra and Trigonometry: A Pre-Calculus Approach,* Third Edition. Englewood Cliffs, NJ: Prentice Hall, 1983, 1991.

* SWOKOWSKI, EARL W. *Algebra and Trigonometry with Analytic Geometry,* Seventh Edition. Boston, MA: PWS-Kent, 1989.

* USISKIN, ZALMAN. *Advanced Algebra with Transformations and Applications.* River Forest, IL: Laidlaw Brothers, 1976.

6.3 Elementary Calculus

* ANTON, HOWARD. *Calculus with Analytic Geometry,* Third Edition. New York, NY: John Wiley, 1980, 1988.

ASH, CAROL AND ASH, ROBERT B. *The Calculus Tutoring Book.* Los Angeles, CA: IEEE Computer Society, 1985.

BERRY, JOHN; NORCLIFFE, ALLAN; AND HUMBLE, STEPHEN. *Introductory Mathematics Through Science Applications.* New York, NY: Cambridge University Press, 1989.

BITTINGER, MARVIN L. *Calculus: A Modeling Approach,* Fourth Edition. Reading, MA: Addison-Wesley, 1980, 1988.

DORN, WILLIAM S.; BITTER, GARY G.; AND HECTOR, DAVID L. *Computer Applications for Calculus.* Boston, MA: Prindle, Weber and Schmidt, 1972.

FEROE, JOHN AND STEINHORN, CHARLES. *Single Variable Calculus with Discrete Mathematics.* San Diego, CA: Harcourt Brace Jovanovich, 1991.

* FINNEY, ROSS L. AND THOMAS, GEORGE B., JR. *Calculus.* Reading, MA: Addison-Wesley, 1990.

FRALEIGH, JOHN B. *Calculus with Analytic Geometry,* Third Edition. Reading, MA: Addison-Wesley, 1990.

* GOLDSTEIN, LARRY J.; LAY, DAVID C.; AND SCHNEIDER, DAVID I. *Calculus and Its Applications,* Fifth Edition. Englewood Cliffs, NJ: Prentice Hall, 1977, 1990.

GROSSMAN, STANLEY I. *Calculus,* Fourth Edition. New York, NY: Academic Press, 1981; San Diego, CA: Harcourt Brace Jovanovich, 1988.

* HAMMING, RICHARD W. *Methods of Mathematics Applied to Calculus, Probability, and Statistics.* Englewood Cliffs, NJ: Prentice Hall, 1985.

HENLE, JAMES M. AND KLEINBERG, EUGENE M. *Infinitesimal Calculus.* Cambridge, MA: MIT Press, 1979.

* KEISLER, H. JEROME. *Elementary Calculus,* Second Edition. Boston, MA: Prindle, Weber and Schmidt, 1976, 1986.

* KEISLER, H. JEROME. *Foundations of Infinitesimal Calculus.* Boston, MA: Prindle, Weber and Schmidt, 1976.

* KLINE, MORRIS. *Calculus: An Intuitive and Physical Approach,* Second Edition. New York, NY: John Wiley, 1977.

* LAX, PETER; BURSTEIN, SAMUEL; AND LAX, ANNELI. *Calculus with Applications and Computing.* New York, NY: Springer-Verlag, 1976.

LEVI, HOWARD. *Polynomials, Power Series, and Calculus.* New York, NY: Van Nostrand Reinhold, 1968.

* MARSDEN, JERROLD E. AND WEINSTEIN, ALAN. *Calculus,* Second Edition. Redwood City, CA: Benjamin Cummings, 1980; New York, NY: Springer-Verlag, 1985.

PRIESTLEY, WILLIAM M. *Calculus: An Historical Approach.* New York, NY: Springer-Verlag, 1979.

SEELEY, ROBERT T. *Calculus.* San Diego, CA: Harcourt Brace Jovanovich, 1990.

* SIMMONS, GEORGE F. *Calculus with Analytic Geometry.* New York, NY: McGraw-Hill, 1985.

SMALL, DONALD B. AND HOSACK, JOHN M. *Calculus: An Integrated Approach.* New York, NY: McGraw-Hill, 1990.

SMALL, DONALD B. AND HOSACK, JOHN M. *Explorations in Calculus with a Computer Algebra System.* New York, NY: McGraw-Hill, 1990.

** SPIVAK, MICHAEL D. *Calculus,* Second Edition. Reading, MA: W.A. Benjamin, 1967; Boston, MA: Publish or Perish, 1980.

** STEIN, SHERMAN K. *Calculus and Analytic Geometry,* Fourth Edition. New York, NY: McGraw-Hill, 1982, 1987.

** STRANG, GILBERT. *Calculus.* Wellesley, MA: Wellesley-Cambridge Press, 1991.

SWOKOWSKI, EARL W. *Calculus,* Fifth Edition. Boston, MA: PWS-Kent, 1975, 1991.

*** THOMAS, GEORGE B., JR. AND FINNEY, ROSS L. *Calculus and Analytic Geometry,* Seventh Edition. Reading, MA: Addison-Wesley, 1968, 1987.

6.4 Advanced Calculus

AMAZIGO, JOHN C. AND RUBENFELD, LESTER A. *Advanced Calculus and Its Applications to the Engineering and Physical Sciences.* New York, NY: John Wiley, 1980.

*** APOSTOL, TOM M. *Calculus,* 2 Vols., Second Edition. New York, NY: John Wiley, 1967, 1969.

BAMBERG, PAUL AND STERNBERG, SHLOMO. *A Course in Mathematics For Students of Physics,* 2 Vols. New York, NY: Cambridge University Press, 1988, 1990.

* BRESSOUD, DAVID M. *Second Year Calculus.* New York, NY: Springer-Verlag, 1991.

BROMWICH, THOMAS J. L'ANSON. *An Introduction to the Theory of Infinite Series.* New York, NY: Macmillan, 1965.

** BUCK, R. CREIGHTON. *Advanced Calculus,* Third Edition. New York, NY: McGraw-Hill, 1965, 1978.

** COURANT, RICHARD AND JOHN, FRITZ. *Introduction to Calculus and Analysis,* 2 Vols. New York, NY: John Wiley, 1965; New York, NY: Springer-Verlag, 1989.

*** COURANT, RICHARD. *Differential and Integral Calculus,* 2 Vols. New York, NY: Interscience, 1937.

CRONIN-SCANLON, JANE. *Advanced Calculus: A Start in Analysis,* Revised Edition. Lexington, MA: D.C. Heath, 1969.

FULKS, WATSON. *Advanced Calculus,* Third Edition. New York, NY: John Wiley, 1969, 1978.

GROSSMAN, STANLEY I. *Multivariable Calculus, Linear Algebra, and Differential Equations,* Second Edition. New York, NY: Academic Press, 1986.

*** HARDY, G.H. *A Course of Pure Mathematics.* New York, NY: Cambridge University Press, 1959.

HILDEBRAND, FRANCIS B. *Advanced Calculus for Applications,* Second Edition. Englewood Cliffs, NJ: Prentice Hall, 1962, 1976.

* KAPLAN, WILFRED. *Advanced Calculus,* Third Edition. Reading, MA: Addison-Wesley, 1952, 1984.

* KNOPP, KONRAD. *Infinite Sequences and Series.* Mineola, NY: Dover, 1956.

* KNOPP, KONRAD. *Theory and Application of Infinite Series.* New York, NY: Hafner Press, 1948; Mineola, NY: Dover, 1990.

LANDAU, EDMUND G.H. *Differential and Integral Calculus.* New York, NY: Chelsea, 1965.

LOOMIS, LYNN H. AND STERNBERG, SHLOMO. *Advanced Calculus,* Revised Edition. Reading, MA: Addison-Wesley, 1968; Boston, MA: Jones and Bartlett, 1990.

MAGNUS, JAN R. AND NEUDECKER, HEINZ. *Matrix Differential Calculus with Applications in Statistics and Econometrics.* New York, NY: John Wiley, 1988.

** MARSDEN, JERROLD E. AND TROMBA, ANTHONY J. *Vector Calculus,* Third Edition. New York, NY: W.H. Freeman, 1976, 1988.

* MARSDEN, JERROLD E. AND WEINSTEIN, ALAN. *Calculus III,* Second Edition. New York, NY: Springer-Verlag, 1985.

PRICE, G. BALEY. *Multivariable Analysis.* New York, NY: Springer-Verlag, 1984.

SAGAN, HANS. *Advanced Calculus of Real-Valued Functions of a Real Variable and Vector-Valued Functions of a Vector Variable.* Boston, MA: Houghton Mifflin, 1974.

** SCHEY, H.M. *Div, Grad, Curl, and All That: An Informal Text on Vector Calculus.* New York, NY: W.W. Norton, 1973.

* SIMMONDS, JAMES G. *A Brief on Tensor Analysis.* New York, NY: Springer-Verlag, 1982.

*** SPIVAK, MICHAEL D. *Calculus on Manifolds.* Reading, MA: W.A. Benjamin, 1965.

TAYLOR, ANGUS E. AND MANN, W. ROBERT. *Advanced Calculus,* Third Edition. New York, NY: John Wiley, 1983.

* WIDDER, DAVID V. *Advanced Calculus,* Second Edition. Englewood Cliffs, NJ: Prentice Hall, 1947; Mineola, NY: Dover, 1987.

WILLIAMSON, RICHARD E.; CROWELL, RICHARD H.; AND TROTTER, HALE F. *Calculus of Vector Functions,* Third Edition. Englewood Cliffs, NJ: Prentice Hall, 1972.

6.5 Supplementary Resources

** APOSTOL, TOM M., ED. *Selected Papers on Calculus.* Washington, DC: Mathematical Association of America, 1969.

** APOSTOL, TOM M., *et al.*, EDS. *Selected Papers on Precalculus.* Washington, DC: Mathematical Association of America, 1977.

BLUMAN, GEORGE W. *Problem Book for First Year Calculus.* New York, NY: Springer-Verlag, 1984.

** CIPRA, BARRY. *Misteaks ... and How to Find Them Before the Teacher Does ...*, Second Edition. New York, NY: Birkhäuser, 1983; New York, NY: Academic Press, 1989.

* DE MESTRE, NEVILLE. *The Mathematics of Projectiles in Sport.* New York, NY: Cambridge University Press, 1990.

* DOBBINS, ROBERT R. *If And Only If in Analysis.* Lanham, MD: University Press of America, 1977.

* GRINSTEIN, LOUISE S. AND MICHAELS, BRENDA, EDS. *Calculus: Readings from the Mathematics Teacher.* Reston, VA: National Council of Teachers of Mathematics, 1977.

* KLAMBAUER, GABRIEL. *Aspects of Calculus.* New York, NY: Springer-Verlag, 1986.

* MAY, KENNETH O., ED. *Lectures on Calculus.* San Francisco, CA: Holden-Day, 1967.

MENDELSON, ELLIOTT. *Schaum's 3000 Solved Problems in Calculus.* New York, NY: McGraw-Hill, 1988.

*** SAWYER, W.W. *What is Calculus About?* Washington, DC: Mathematical Association of America, 1975.

* SWANN, HOWARD AND JOHNSON, JOHN. *Prof. E. McSquared's Expanded Intergalactic Version: A Calculus Primer.* Providence, RI: Janson, 1989.

** TOEPLITZ, OTTO. *Calculus: A Genetic Approach.* Chicago, IL: University of Chicago Press, 1963.

7 Differential Equations

7.1 Introductory Texts

*** BOYCE, WILLIAM E. AND DIPRIMA, RICHARD C. *Elementary Differential Equations and Boundary Value Problems,* Fifth Edition. New York, NY: John Wiley, 1969, 1992.

* BRAUN, MARTIN. *Differential Equations and Their Applications: An Introduction to Applied Mathematics,* Third Edition. New York, NY: Springer-Verlag, 1975, 1983.

BURGHES, DAVID N. AND BORRIE, M.S. *Modelling with Differential Equations.* New York, NY: Halsted Press, 1981.

* CODDINGTON, EARL A. *An Introduction to Ordinary Differential Equations.* Englewood Cliffs, NJ: Prentice Hall, 1961; Mineola, NY: Dover, 1989.

** EDWARDS, C.H., JR. AND PENNEY, DAVID E. *Elementary Differential Equations with Applications,* Second Edition. Englewood Cliffs, NJ: Prentice Hall, 1985, 1989.

HOCHSTADT, HARRY. *Differential Equations.* New York, NY: Holt, Rinehart and Winston, 1964; Mineola, NY: Dover, 1975.

** HUBBARD, JOHN H. AND WEST, BEVERLY H. *Differential Equations: A Dynamical Systems Approach,* New York, NY: Springer-Verlag, 1991.

MILLER, RICHARD K. *Ordinary Differential Equations.* New York, NY: Academic Press, 1982.

** REDHEFFER, RAY AND PORT, DAN. *Differential Equations: Theory and Applications.* Boston, MA: Jones and Bartlett, 1991.

ROBERTS, CHARLES E., JR. *Ordinary Differential Equations: A Computational Approach.* Englewood Cliffs, NJ: Prentice Hall, 1979.

** SIMMONS, GEORGE F. AND ROBERTSON, JOHN S. *Differential Equations with Applications and Historical Notes,* Second Edition. New York, NY: McGraw-Hill, 1972, 1991.

** ZILL, DENNIS G. *A First Course in Differential Equations with Applications,* Fourth Edition. Boston, MA: PWS-Kent, 1980, 1989.

7.2 Ordinary Differential Equations

** ARNOLD, V.I. *Geometrical Methods in the Theory of Ordinary Differential Equations,* Second Edition. New York, NY: Springer-Verlag, 1983, 1988.

* ARNOLD, V.I. *Ordinary Differential Equations.* Cambridge, MA: MIT Press, 1973, 1978.

BELLMAN, RICHARD E. *Stability Theory of Differential Equations.* Mineola, NY: Dover, 1969.

*** BIRKHOFF, GARRETT AND ROTA, GIAN-CARLO. *Ordinary Differential Equations,* Fourth Edition. New York, NY: John Wiley, 1969, 1989.

** BRAUER, FRED AND NOHEL, JOHN A. *The Qualitative Theory of Ordinary Differential Equations: An Introduction.* Reading, MA: W.A. Benjamin, 1969; Mineola, NY: Dover, 1989.

CARRIER, GEORGE F. AND PEARSON, CARL E. *Ordinary Differential Equations.* New York, NY: Blaisdell, 1968; Philadelphia, PA: Society for Industrial and Applied Mathematics, 1991.

CESARI, LAMBERTO. *Asymptotic Behavior and Stability Problems in Ordinary Differential Equations.* New York, NY: Springer-Verlag, 1963.

*** CODDINGTON, EARL A. AND LEVINSON, NORMAN. *Theory of Ordinary Differential Equations.* New York, NY: McGraw-Hill, 1955; Melbourne, FL: Robert E. Krieger, 1984.

DRIVER, RODNEY D. *Ordinary and Delay Differential Equations.* New York, NY: Springer-Verlag, 1977.

HALE, JACK K., ED. *Studies in Ordinary Differential Equations.* Washington, DC: Mathematical Association of America, 1977.

* HALE, JACK K. *Ordinary Differential Equations.* New York, NY: John Wiley, 1969; Melbourne, FL: Robert E. Krieger, 1980.

* HARTMAN, PHILIP. *Ordinary Differential Equations,* Second Edition. New York, NY: Birkhäuser, 1973, 1982.

HILLE, EINAR. *Ordinary Differential Equations in the Complex Domain.* New York, NY: John Wiley, 1976.

*** HIRSCH, MORRIS W. AND SMALE, STEPHEN. *Differential Equations, Dynamical Systems, and Linear Algebra.* New York, NY: Academic Press, 1974.

HUREWICZ, WITOLD. *Lectures on Ordinary Differential Equations.* Cambridge, MA: MIT Press, 1958; Mineola, NY: Dover, 1990.

* INCE, EDWARD L. *Ordinary Differential Equations.* Mineola, NY: Dover, 1956.

JORDAN, D. AND SMITH, P. *Nonlinear Ordinary Differential Equations,* Second Edition. New York, NY: Clarendon Press, 1987.

LAKIN, WILLIAM D. AND SANCHEZ, DAVID A. *Topics in Ordinary Differential Equations.* Mineola, NY: Dover, 1982.

* LEFSCHETZ, SOLOMON. *Differential Equations: Geometric Theory.* Mineola, NY: Dover, 1977.

NEMYTSKII, V.V. AND STEPANOV, V.V. *Qualitative Theory of Differential Equations.* Princeton, NJ: Princeton University Press, 1960; Mineola, NY: Dover, 1989.

PONTRJAGIN, LEV S. *Ordinary Differential Equations.* Reading, MA: Addison-Wesley, 1962.

REID, WILLIAM T. *Sturmian Theory for Ordinary Differential Equations.* New York, NY: Springer-Verlag, 1980.

* SANCHEZ, DAVID A. *Ordinary Differential Equations and Stability Theory: An Introduction.* Mineola, NY: Dover, 1979.

STRUBLE, RAIMOND A. *Nonlinear Differential Equations.* Melbourne, FL: Robert E. Krieger, 1983.

** WALTMAN, PAUL. *A Second Course in Elementary Differential Equations.* New York, NY: Academic Press, 1986.

WASOW, WOLFGANG. *Asymptotic Expansions for Ordinary Differential Equations.* Mineola, NY: Dover, 1987.

7.3 Dynamical Systems

* ABRAHAM, RALPH H. AND SHAW, CHRISTOPHER D., EDS. *Dynamics—The Geometry of Behavior,* 4 Vols. Santa Cruz, CA: Aerial Press, 1982–88.

* ARROWSMITH, D.K. AND PLACE, C.M. *An Introduction to Dynamical Systems.* New York, NY: Cambridge University Press, 1990.

BHATIA, NAM PARSHAD. *Stability Theory of Dynamical Systems.* New York, NY: Springer-Verlag, 1970.

BURTON, T.A. *Stability and Periodic Solutions of Ordinary and Functional Differential Equations.* New York, NY: Academic Press, 1985.

CHOW, S. AND HALE, JACK K. *Methods of Bifurcation Theory.* New York, NY: Springer-Verlag, 1982.

*** GUCKENHEIMER, JOHN AND HOLMES, PHILIP. *Nonlinear Oscillations, Dynamical Systems, and Bifurcations of Vector Fields.* New York, NY: Springer-Verlag, 1983.

HALE, JACK K.; MAGALHAES, LUIS T.; AND OLIVA, WALDYR M. *An Introduction to Infinite Dimensional Dynamical Systems—Geometric Theory.* New York, NY: Springer-Verlag, 1984.

HAO, BAI-LIN. *Chaos.* Teaneck, NJ: World Scientific, 1984.

HAZEWINKEL, M.; JURKOVICH, R.; AND PAELINCK, J.H.P., EDS. *Bifurcation Analysis: Principles, Applications, and Synthesis.* Norwell, MA: D. Reidel, 1985.

* IOOSS, GÉRARD AND JOSEPH, DANIEL D. *Elementary Stability and Bifurcation Theory,* Second Edition. New York, NY: Springer-Verlag, 1990.

LASALLE, JOSEPH P. AND LEFSCHETZ, SOLOMON. *Stability by Liapunov's Direct Method.* New York, NY: Academic Press, 1961.

LASALLE, JOSEPH P. *The Stability of Dynamical Systems.* Philadelphia, PA: Society for Industrial and Applied Mathematics, 1976.

LIAPUNOV, A. *Stability of Motion.* New York, NY: Academic Press, 1966.

* Marsden, Jerrold E. *The Hopf Bifurcation and Its Applications.* New York, NY: Springer-Verlag, 1976.

Percival, Ian and Richards, Derek. *Introduction to Dynamics.* New York, NY: Cambridge University Press, 1982.

Rouche, N.; Habets, P.; and Laloy, M. *Stability Theory by Liapunov's Direct Method.* New York, NY: Springer-Verlag, 1977.

** Ruelle, David. *Elements of Differentiable Dynamics and Bifurcation Theory.* New York, NY: Academic Press, 1989.

* Sparrow, Colin. *The Lorenz Equations: Bifurcations, Chaos, and Strange Attractors.* New York, NY: Springer-Verlag, 1982.

Verhulst, F. *Nonlinear Differential Equations and Dynamical Systems,* Second Edition. New York, NY: Springer-Verlag, 1985, 1990.

** Wiggins, Stephen. *Introduction to Applied Nonlinear Dynamical Systems and Chaos.* New York, NY: Springer-Verlag, 1990.

7.4 Partial Differential Equations

Andrews, Larry C. *Elementary Partial Differential Equations with Boundary Value Problems.* New York, NY: Academic Press, 1986.

Baker, Bevan B. and Copson, E.T. *The Mathematical Theory of Huygens' Principle.* New York, NY: Chelsea, 1987.

* Bergman, Stefan. *Kernel Functions and Elliptic Differential Equations in Mathematical Physics.* New York, NY: Academic Press, 1953.

Bleistein, Norman. *Mathematical Methods for Wave Phenomena.* New York, NY: Academic Press, 1984.

** Carrier, George F. and Pearson, Carl E. *Partial Differential Equations: Theory and Technique,* Second Edition. New York, NY: Academic Press, 1988.

Farlow, Stanley J. *Partial Differential Equations for Scientists and Engineers.* New York, NY: John Wiley, 1982.

Friedman, Avner. *Partial Differential Equations.* New York, NY: Holt, Rinehart and Winston, 1969.

** Garabedian, Paul R. *Partial Differential Equations,* Second Edition. New York, NY: John Wiley, 1964; New York, NY: Chelsea, 1986.

* Gustafson, Karl E. *Introduction to Partial Differential Equations and Hilbert Space Methods,* Second Edition. New York, NY: John Wiley, 1980, 1987.

* Haberman, Richard. *Elementary Applied Partial Differential Equations with Fourier Series and Boundary Value Problems,* Second Edition. Englewood Cliffs, NJ: Prentice Hall, 1983, 1987.

Hellwig, Gunter. *Partial Differential Equations.* New York, NY: Blaisdell, 1964.

* Hörmander, Lars. *The Analysis of Linear Partial Differential Operators I: Distribution Theory and Fourier Analysis,* Second Edition. New York, NY: Springer-Verlag, 1983, 1990.

John, Fritz. *Partial Differential Equations,* Fourth Edition. New York, NY: Springer-Verlag, 1971, 1982.

Littman, Walter, ed. *Studies in Partial Differential Equations.* Washington, DC: Mathematical Association of America, 1982.

Pinsky, Mark A. *Introduction to Partial Differential Equations with Applications,* Second Edition. New York, NY: McGraw-Hill, 1984, 1991.

* Protter, Murray H. and Weinberger, Hans F. *Maximum Principles in Differential Equations.* Englewood Cliffs, NJ: Prentice Hall, 1967.

Smoller, J. *Shock Waves and Reaction Diffusion Equations.* New York, NY: Springer-Verlag, 1983.

Tikhonov, Andrei N. *Partial Differential Equations of Mathematical Physics.* San Francisco, CA: Holden-Day, 1967.

TREVES, FRANÇOIS. *Basic Linear Partial Differential Equations.* New York, NY: Academic Press, 1975.

** WEINBERGER, HANS F. *A First Course in Partial Differential Equations with Complex Variables and Transform Methods.* Lexington, MA: Xerox College, 1965.

* WHITHAM, G. *Linear and Nonlinear Waves.* New York, NY: John Wiley, 1974.

** WIDDER, DAVID V. *The Heat Equation.* New York, NY: Academic Press, 1975.

WILLIAMS, W.E. *Partial Differential Equations.* New York, NY: Clarendon Press, 1980.

ZACHMANOGLOU, E.C. AND THOE, DALE W. *Introduction to Partial Differential Equations with Applications.* Baltimore, MD: Williams and Wilkins, 1976; Mineola, NY: Dover, 1986.

* ZAUDERER, ERICH. *Partial Differential Equations of Applied Mathematics,* Second Edition. New York, NY: John Wiley, 1983, 1989.

7.5 Boundary Value Problems

*** CHURCHILL, RUEL V. AND BROWN, JAMES W. *Fourier Series and Boundary Value Problems,* Third Edition. New York, NY: McGraw-Hill, 1978.

CRANK, JOHN. *Free and Moving Boundary Problems.* New York, NY: Clarendon Press, 1984.

GREENBERG, MICHAEL D. *Application of Green's Functions in Science and Engineering.* Englewood Cliffs, NJ: Prentice Hall, 1971.

HANNA, J. RAY AND ROWLAND, J.H. *Fourier Series and Integrals of Boundary Value Problems,* Second Edition. New York, NY: John Wiley, 1982, 1990.

* POWERS, DAVID L. *Boundary Value Problems,* Third Edition. New York, NY: Academic Press, 1979, 1987.

ROACH, G.F. *Green's Functions,* Second Edition. New York, NY: Cambridge University Press, 1982.

** STAKGOLD, IVAR. *Green's Functions and Boundary Value Problems.* New York, NY: John Wiley, 1979.

7.6 Special Topics

* GOULD, SYDNEY H. *Variational Methods for Eigenvalue Problems: An Introduction to the Method.* Toronto: University of Toronto Press, 1957, 1966.

HOCHSTADT, HARRY. *Integral Equations.* New York, NY: John Wiley, 1973.

LADDE, G.S. *Oscillation Theory of Differential Equations with Deviating Arguments.* New York, NY: Marcel Dekker, 1987.

MARTIN, ROBERT H., JR. *Nonlinear Operators and Differential Equations in Banach Spaces.* New York, NY: John Wiley, 1976; Melbourne, FL: Robert E. Krieger, 1987.

* MICKENS, RONALD E. *An Introduction to Nonlinear Oscillations.* New York, NY: Cambridge University Press, 1981.

MILLER, RICHARD K. *Nonlinear Volterra Integral Equations.* Reading, MA: W.A. Benjamin, 1971.

SMITH, DONALD R. *Singular-perturbation Theory: An Introduction with Applications.* New York, NY: Cambridge University Press, 1985.

TIKHONOV, ANDREI N. AND SAMARSKII, A.A. *Equations of Mathematical Physics.* New York, NY: Macmillan, 1963; Mineola, NY: Dover, 1990.

TITCHMARSH, EDWARD C. *Eigenfunction Expansions Associated with Second-Order Differential Equations.* New York, NY: Clarendon Press, 1962.

* TRICOMI, FRANCESCO G. *Integral Equations.* Mineola, NY: Dover, 1985.

VOLTERRA, VITO. *Theory of Functionals and of Integral and Integro-Differential Equations.* Mineola, NY: Dover, 1959.

* YOSIDA, KÔSAKU. *Lectures on Differential and Integral Equations.* New York, NY: Interscience, 1960.

8 Analysis

8.1 Foundations of Analysis

BARNIER, WILLIAM AND FELDMAN, NORMAN. *Introduction to Advanced Mathematics.* Englewood Cliffs, NJ: Prentice Hall, 1990.

BITTINGER, MARVIN L. *Logic, Proof, and Sets,* Second Edition. Reading, MA: Addison-Wesley, 1982.

FEFERMAN, SOLOMON. *The Number Systems: Foundations of Algebra and Analysis,* Second Edition. Reading, MA: Addison-Wesley, 1964; New York, NY: Chelsea, 1989.

* FENDEL, DANIEL AND RESEK, DIANE. *Foundations of Higher Mathematics: Exploration and Proof.* Reading, MA: Addison-Wesley, 1990.

* GARDINER, A. *Infinite Processes: Background to Analysis.* New York, NY: Springer-Verlag, 1982.

GLEASON, ANDREW M. *Fundamentals of Abstract Analysis.* Reading, MA: Addison-Wesley, 1966; Boston, MA: Jones and Bartlett, 1991.

** LANDAU, EDMUND G.H. *The Foundations of Analysis,* Third Edition. New York, NY: Chelsea, 1951, 1966.

LIGHTSTONE, A.H. *Symbolic Logic and the Real Number System: An Introduction to the Foundations of Number Systems.* New York, NY: Harper and Row, 1965.

LUCAS, JOHN F. *Introduction to Abstract Mathematics,* Second Edition. Belmont, CA: Wadsworth, 1986; New York, NY: Ardsley House, 1990.

SMITH, D.; EGGEN, M.; AND ST. ANDRE, R. *A Transition to Advanced Mathematics,* Third Edition. Pacific Grove, CA: Brooks/Cole, 1990.

** SOLOW, DANIEL. *How To Read and Do Proofs: An Introduction to Mathematical Thought Processes,* Second Edition. New York, NY: John Wiley, 1982, 1990.

8.2 Elementary Real Analysis

ALIPRANTIS, CHARALAMBOS D. AND BURKINSHAW, OWEN. *Principles of Real Analysis.* New York, NY: Elsevier Science, 1981.

* APOSTOL, TOM M. *Mathematical Analysis,* Second Edition. Reading, MA: Addison-Wesley, 1974.

* BARTLE, ROBERT G. *Elements of Real Analysis,* Second Edition. New York, NY: John Wiley, 1976.

BINMORE, K.G. *Mathematical Analysis: A Straightforward Approach,* 2 Vols. New York, NY: Cambridge University Press, 1977, 1981.

BURKILL, JOHN C. *A First Course in Mathematical Analysis.* New York, NY: Cambridge University Press, 1978.

GOFFMAN, CASPER. *Real Functions,* Revised Edition. Boston, MA: Prindle, Weber and Schmidt, 1967.

* GOLDBERG, RICHARD R. *Methods of Real Analysis,* Second Edition. New York, NY: John Wiley, 1976.

KOLMOGOROV, ANDREI N. AND FOMIN, S.V. *Introductory Real Analysis.* Mineola, NY: Dover, 1975.

PROTTER, MURRAY H. AND MORREY, C.B. *A First Course in Real Analysis.* New York, NY: Springer-Verlag, 1977.

ROSENLICHT, MAXWELL. *Introduction to Analysis.* Mineola, NY: Dover, 1986.

** ROSS, KENNETH A. *Elementary Analysis: The Theory of Calculus.* New York, NY: Springer-Verlag, 1980.

** ROYDEN, H.L. *Real Analysis,* Third Edition. New York, NY: Macmillan, 1968, 1988.

*** RUDIN, WALTER. *Principles of Mathematical Analysis,* Third Edition. New York, NY: McGraw-Hill, 1953, 1976.

SMITH, KENNAN T. *Primer of Modern Analysis.* Belmont, CA: Bogden and Quigley, 1971; New York, NY: Springer-Verlag, 1983.

WHEEDEN, RICHARD L. AND ZYGMUND, ANTONI. *Measure and Integral: An Introduction to Real Analysis.* New York, NY: Marcel Dekker, 1977.

8.3 Advanced Real Analysis

AKHIEZER, N.I. *The Classical Moment Problem.* New York, NY: Hafner Press, 1965.

BEALS, R. *Advanced Mathematical Analysis.* New York, NY: Springer-Verlag, 1973.

* BISHOP, ERRETT AND BRIDGES, DOUGLAS S. *Constructive Analysis.* New York, NY: Springer-Verlag, 1985.

*** BOAS, RALPH P., JR. *A Primer of Real Functions,* Third Edition. Washington, DC: Mathematical Association of America, 1972, 1981.

BURKILL, JOHN C. *The Lebesgue Integral.* New York, NY: Cambridge University Press, 1951.

CARATHÉODORY, C. *Algebraic Theory of Measure and Integration,* Second English Edition. New York, NY: Chelsea, 1986.

FISCHER, EMANUEL. *Intermediate Real Analysis.* New York, NY: Springer-Verlag, 1983.

*** GELBAUM, BERNARD R. AND OLMSTED, JOHN M.H. *Theorems and Counterexamples in Mathematics.* (Former title: *Counterexamples in Analysis.*) San Francisco, CA: Holden-Day, 1964; New York, NY: Springer-Verlag, 1990.

*** GILLMAN, LEONARD AND JERISON, MEYER. *Rings of Continuous Functions.* New York, NY: Springer-Verlag, 1976.

GRENANDER, ULF AND SZEGŐ, GÁBOR. *Toeplitz Forms and Their Applications.* New York, NY: Chelsea, 1984.

** HALMOS, PAUL R. *Measure Theory.* New York, NY: Springer-Verlag, 1974.

HARDY, G.H. *Divergent Series.* New York, NY: Oxford University Press, 1949.

* HEWITT, EDWIN AND STROMBERG, KARL R. *Real and Abstract Analysis: A Modern Treatment of the Theory of Functions of a Real Variable.* New York, NY: Springer-Verlag, 1969, 1975.

KARLIN, SAMUEL AND STUDDEN, W. *Tchebycheff Systems.* New York, NY: John Wiley, 1966.

MOISE, EDWIN E. *Introductory Problem Courses in Analysis and Topology.* New York, NY: Springer-Verlag, 1982.

MUNKRES, JAMES R. *Analysis on Manifolds.* Reading, MA: Addison-Wesley, 1991.

MUNROE, MARSHALL E. *Introduction to Measure and Integration,* Second Edition. Reading, MA: Addison-Wesley, 1971.

*** PÓLYA, GEORGE AND SZEGŐ, GÁBOR. *Problems and Theorems in Analysis,* 2 Vols. New York, NY: Springer-Verlag, 1972, 1976.

STROMBERG, KARL R. *Introduction to Classical Real Analysis.* Belmont, CA: Wadsworth, 1981.

TAYLOR, ANGUS E. *General Theory of Functions and Integration.* Mineola, NY: Dover, 1985.

* WAGON, STAN. *The Banach-Tarski Paradox.* New York, NY: Cambridge University Press, 1985.

8.4 Fourier Analysis

CHANDRASEKHARAN, KOMARAVOLU. *Classical Fourier Transforms.* New York, NY: Springer-Verlag, 1989.

CHIHARA, T.S. *An Introduction to Orthogonal Polynomials.* New York, NY: Gordon and Breach, 1978.

DAVIES, B. *Integral Transforms and their Applications,* Second Edition. New York, NY: Springer-Verlag, 1978, 1985.

* DYM, H. AND MCKEAN, H. *Fourier Series and Integrals.* New York, NY: Academic Press, 1972.

EDWARDS, ROBERT E. *Fourier Series: A Modern Introduction,* 2 Vols., Second Edition. New York, NY: Springer-Verlag, 1979, 1982.

JACKSON, DUNHAM. *Fourier Series and Orthogonal Polynomials.* Washington, DC: Mathematical Association of America, 1941.

* KÖRNER, T.W. *Fourier Analysis.* New York, NY: Cambridge University Press, 1988.

ROGOSINSKI, WERNER. *Fourier Series,* Second Edition. New York, NY: Chelsea, 1962.

SEELEY, ROBERT T. *An Introduction to Fourier Series and Integrals.* Reading, MA: W.A. Benjamin, 1966.

** STEIN, E.M. AND WEISS, G. *Introduction to Fourier Analysis on Euclidean Spaces.* Princeton, NJ: Princeton University Press, 1971.

* SZEGŐ, GÁBOR. *Orthogonal Polynomials,* Fourth Edition. Providence, RI: American Mathematical Society, 1975.

* TITCHMARSH, EDWARD C. *Introduction to the Theory of Fourier Integrals.* London: Oxford University Press, 1948.

WIDDER, DAVID V. *An Introduction to Transform Theory.* New York, NY: Academic Press, 1971.

** WIENER, NORBERT. *The Fourier Integral and Certain of its Applications.* New York, NY: Cambridge University Press, 1933, 1988.

*** ZYGMUND, ANTONI. *Trigonometric Series.* New York, NY: Cambridge University Press, 1968, 1988.

8.5 Fractals

* BARNSLEY, MICHAEL. *Fractals Everywhere.* New York, NY: Academic Press, 1988.

DEVANEY, ROBERT L. AND KEEN, LINDA, EDS. *Chaos and Fractals: The Mathematics Behind the Computer Graphics.* Providence, RI: American Mathematical Society, 1989.

** DEVANEY, ROBERT L. *An Introduction to Chaotic Dynamical Systems,* Second Edition. Redwood City, CA: Benjamin Cummings, 1986, 1989.

* DEVANEY, ROBERT L. *Chaos, Fractals, and Dynamics: Computer Experiments in Mathematics.* Reading, MA: Addison-Wesley, 1990.

EDGAR, G.A. *Measure, Topology, and Fractal Geometry.* New York, NY: Springer-Verlag, 1990.

FALCONER, KENNETH J. *The Geometry of Fractal Sets.* New York, NY: Cambridge University Press, 1985, 1986.

* LAUWERIER, HANS. *Fractals: Endlessly Repeated Geometrical Figures.* Princeton, NJ: Princeton University Press, 1991.

*** MANDELBROT, BENOIT. *The Fractal Geometry of Nature.* New York, NY: W.H. Freeman, 1982.

* PEITGEN, HEINZ-OTTO AND RICHTER, P.H. *The Beauty of Fractals: Images of Complex Dynamical Systems.* New York, NY: Springer-Verlag, 1986.

** PEITGEN, HEINZ-OTTO AND SAUPE, DIETMAR, EDS. *The Science of Fractal Images.* New York, NY: Springer-Verlag, 1988.

PRESTON, CHRIS. *Iterates of Maps on an Interval.* New York, NY: Springer-Verlag, 1983.

** SCHROEDER, MANFRED R. *Fractals, Chaos, Power Laws: Minutes from an Infinite Paradise.* New York, NY: W.H. Freeman, 1990.

8.6 Introductory Complex Analysis

*** AHLFORS, LARS V. *Complex Analysis: An Introduction to the Theory of Analytic Functions of One Complex Variable,* Third Edition. New York, NY: McGraw-Hill, 1966, 1979.

BAK, JOSEPH AND NEWMAN, DONALD J. *Complex Analysis.* New York, NY: Springer-Verlag, 1982.

* BOAS, RALPH P., JR. *Invitation to Complex Analysis.* New York, NY: Birkhäuser, 1987.

BURCKEL, ROBERT B. *An Introduction to Classical Complex Analysis.* New York, NY: Academic Press, 1979.

CARTAN, HENRI. *Theory of Analytic Functions of One or Several Complex Variables.* Reading, MA: Addison-Wesley, 1983.

** CONWAY, JOHN B. *Functions of One Complex Variable,* Second Edition. New York, NY: Springer-Verlag, 1973, 1978.

HEINS, MAURICE. *Complex Function Theory.* New York, NY: Academic Press, 1968.

* KNOPP, KONRAD. *Theory of Functions,* 2 Vols. Mineola, NY: Dover, 1945, 1947; 1968.

MARKUSHEVICH, A.I. *The Theory of Analytic Functions: A Brief Course.* Moscow: MIR, 1983.

NARASIMHAN, R. *Complex Analysis in One Variable.* New York, NY: Birkhäuser, 1985.

NEHARI, ZEEV. *Introduction to Complex Analysis,* Revised Edition. Boston, MA: Allyn and Bacon, 1968.

NEVANLINNA, ROLF AND PAATERO, V. *Introduction to Complex Analysis.* Reading, MA: Addison-Wesley, 1969.

* PÓLYA, GEORGE AND LATTA, GORDON. *Complex Variables.* New York, NY: John Wiley, 1974.

8.7 Advanced Complex Analysis

* CARATHÉODORY, C. *Theory of Functions of a Complex Variable,* 2 Vols. New York, NY: Chelsea, 1958.

DAVIS, PHILIP J. *The Schwarz Function and Its Applications.* Washington, DC: Mathematical Association of America, 1974.

FISHER, STEPHEN D. *Complex Variables,* Second Edition. Belmont, CA: Wadsworth, 1986, 1990.

GRAUERT, H. AND FRITZSCHE, K. *Several Complex Variables.* New York, NY: Springer-Verlag, 1976.

** HENRICI, PETER. *Applied and Computational Complex Analysis,* 3 Vols. New York, NY: John Wiley, 1974–86.

* HILLE, EINAR. *Analytic Function Theory,* 2 Vols. New York, NY: Chelsea, 1973.

JONES, WILLIAM B. AND THRON, WOLFGANG J. *Continued Fractions: Analytic Theory and Applications.* Reading, MA: Addison-Wesley, 1980.

*** KRANTZ, STEVEN G. *Complex Analysis: The Geometric Viewpoint.* Washington, DC: Mathematical Association of America, 1990.

* LITTLEWOOD, J.E. *Some Problems in Real and Complex Analysis.* Lexington, MA: D.C. Heath, 1968.

* NEHARI, ZEEV. *Conformal Mapping.* New York, NY: McGraw-Hill, 1952; Mineola, NY: Dover, 1979.

PRICE, G. BALEY. *An Introduction to Multicomplex Spaces and Functions.* New York, NY: Marcel Dekker, 1991.

* REMMERT, REINHOLD. *Theory of Complex Functions.* New York, NY: Springer-Verlag, 1991.

*** RUDIN, WALTER. *Real and Complex Analysis,* Third Edition. New York, NY: McGraw-Hill, 1974, 1987.

** SAKS, S. AND ZYGMUND, ANTONI. *Analytic Functions,* Third Edition. New York, NY: American Elsevier, 1952, 1971.

SMITH, KENNAN T. *Power Series from a Computational Point of View.* New York, NY: Springer-Verlag, 1987.

* SPRINGER, GEORGE M. *Introduction to Riemann Surfaces.* Reading, MA: Addison-Wesley, 1957.

** TITCHMARSH, EDWARD C. *Theory of Functions,* Second Edition. New York, NY: Oxford University Press, 1939.

WERMER, JOHN. *Banach Algebras and Several Complex Variables,* Second Edition. New York, NY: Springer-Verlag, 1976.

* WEYL, HERMANN. *The Concept of a Riemann Surface.* Reading, MA: Addison-Wesley, 1964.

*** WHITTAKER, EDMUND T. AND WATSON, G.N. *A Course of Modern Analysis,* Fourth Edition. New York, NY: Cambridge University Press, 1958, 1963.

8.8 Functional Analysis

BEAUZAMY, BERNARD. *Introduction to Banach Spaces and Their Geometry.* Amsterdam: North-Holland, 1982.

BERBERIAN, STERLING K. *Introduction to Hilbert Space,* Second Edition. New York, NY: Chelsea, 1976.

BOLLOBÁS, BÉLA. *Linear Analysis.* New York, NY: Cambridge University Press, 1990.

BRIDGES, DOUGLAS S. *Constructive Functional Analysis.* Brooklyn, NY: Pitman, 1979.

DIEUDONNÉ, JEAN. *Treatise on Analysis,* 7 Vols. New York, NY: Academic Press, 1969–88.

* DUNFORD, NELSON AND SCHWARTZ, JACOB T. *Linear Operators,* Parts I and II. New York, NY: John Wiley, 1958, 1963.

FLETT, T.M. *Differential Analysis: Differentiation, Differential Equations, and Differential Inequalities.* New York, NY: Cambridge University Press, 1980.

GAMELIN, THEODORE W. *Uniform Algebras.* Englewood Cliffs, NJ: Prentice Hall, 1969; New York, NY: Chelsea, 1984.

* GEL'FAND, ISRAEL M., *et al. Generalized Functions,* 5 Vols. New York, NY: Academic Press, 1964–68.

* GEL'FAND, ISRAEL M.; RAIKOV, D.A.; AND SHILOV, G.E. *Commutative Normed Rings.* New York, NY: Chelsea, 1964.

GOFFMAN, CASPER AND PEDRICK, GEORGE. *A First Course in Functional Analysis,* Second Edition. Englewood Cliffs, NJ: Prentice Hall, 1965; New York, NY: Chelsea, 1983.

GROTHENDIECK, A. *Topological Vector Spaces.* New York, NY: Gordon and Breach, 1973.

HALMOS, PAUL R. *A Hilbert Space Problem Book,* Second Edition. New York, NY: Van Nostrand Reinhold, 1967; New York, NY: Springer-Verlag, 1982.

HALMOS, PAUL R. *Introduction to Hilbert Space.* New York, NY: Chelsea, 1951.

* HILLE, EINAR AND PHILLIPS, R.S. *Functional Analysis and Semi-Groups.* Providence, RI: American Mathematical Society, 1957.

* HOFFMAN, KENNETH. *Banach Spaces of Analytic Functions.* Englewood Cliffs, NJ: Prentice Hall, 1962; Mineola, NY: Dover, 1980.

KIRILLOV, A.A. AND GVISHIANI, A.D. *Theorems and Problems in Functional Analysis.* New York, NY: Springer-Verlag, 1982.

* LIUSTERNIK, L. AND SOBOLEV, V. *Elements of Functional Analysis.* New York, NY: Frederick Ungar, 1961.

LORCH, EDGAR R. *Spectral Theory.* New York, NY: Oxford University Press, 1962.

NACHBIN, LEOPOLDO. *Introduction to Functional Analysis: Banach Spaces and Differential Calculus.* New York, NY: Marcel Dekker, 1981.

* NAIMARK, M.A. *Normed Rings.* Groningen: Wolters-Noordhoff, 1960.

** RIESZ, FRIGYES AND NAGY, BELA SZ. *Functional Analysis.* New York, NY: Frederick Ungar, 1955; Mineola, NY: Dover, 1990.

** RUDIN, WALTER. *Functional Analysis.* New York, NY: McGraw-Hill, 1973.

* TAYLOR, ANGUS E. AND LAY, DAVID C. *Introduction to Functional Analysis,* Second Edition. New York, NY: John Wiley, 1958, 1980.

** YOSIDA, KÔSAKU. *Functional Analysis,* Sixth Edition. New York, NY: Springer-Verlag, 1965, 1980.

YOUNG, NICHOLAS. *An Introduction to Hilbert Space.* New York, NY: Cambridge University Press, 1988.

8.9 Operator Theory

ARVESON, WILLIAM. *An Invitation to C^*-Algebras.* New York, NY: Springer-Verlag, 1976.

** BANACH, STEFAN. *Theory of Linear Operators.* Warsaw: 1932; New York, NY: Elsevier Science, 1987.

BROWN, ARLEN AND PEARCY, CARL. *Introduction to Operator Theory I: Elements of Functional Analysis.* New York, NY: Springer-Verlag, 1977.

GOHBERG, ISRAEL AND GOLDBERG, SEYMOUR. *Basic Operator Theory.* New York, NY: Birkhäuser, 1981.

* KADISON, RICHARD V. AND RINGROSE, JOHN R. *Fundamentals of the Theory of Operator Algebras,* Vol. I: Elementary Theory. New York, NY: Academic Press, 1983.

* KATO, TOSIO. *Perturbation Theory for Linear Operators,* Second Edition. New York, NY: Springer-Verlag, 1976.

8.10 Calculus of Variations

IOFFE, A.D. AND TIHOMIROV, V.M. *Theory of Extremal Problems.* Amsterdam: North-Holland, 1979.

KRASNOV, M.L.; MAKARENKO, G.I.; AND KISELYOV, A.I. *Problems and Exercises in the Calculus of Variations.* Moscow: MIR, 1984.

* TROUTMAN, JOHN L. AND HRUSA, W. *Variational Calculus with Elementary Convexity.* New York, NY: Springer-Verlag, 1983.

WEINSTOCK, ROBERT. *Calculus of Variations with Applications to Physics and Engineering.* Mineola, NY: Dover, 1974.

8.11 Inequalities

** BECKENBACH, EDWIN F. AND BELLMAN, RICHARD E. *An Introduction to Inequalities.* Washington, DC: Mathematical Association of America, 1975.

BECKENBACH, EDWIN F. AND BELLMAN, RICHARD E. *Inequalities,* Second Edition. New York, NY: Springer-Verlag, 1961, 1965.

*** HARDY, G.H.; LITTLEWOOD, J.E.; AND PÓLYA, GEORGE. *Inequalities,* Second Edition. New York, NY: Cambridge University Press, 1952, 1988.

* KAZARINOFF, NICHOLAS D. *Analytic Inequalities.* New York, NY: Holt, Rinehart and Winston, 1961.

KOROVKIN, P.P. *Inequalities.* Moscow: MIR, 1975, 1986.

MARSHALL, ALBERT W. AND OLKIN, INGRAM. *Inequalities: Theory of Majorization and Its Applications.* New York, NY: Academic Press, 1979.

TONG, Y. L. *Probability Inequalities in Multivariate Distributions.* New York, NY: Academic Press, 1980.

8.12 Harmonic Analysis

** ASH, J.M., ED. *Studies in Harmonic Analysis.* Washington, DC: Mathematical Association of America, 1976.

HELSON, HENRY. *Harmonic Analysis.* Belmont, CA: Wadsworth, 1991.

* KATZNELSON, YITZHAK. *An Introduction to Harmonic Analysis,* Second Edition. New York, NY: John Wiley, 1968; Mineola, NY: Dover, 1976.

LOOMIS, LYNN H. *An Introduction to Abstract Harmonic Analysis.* New York, NY: Van Nostrand Reinhold, 1953.

* RUDIN, WALTER. *Fourier Analysis on Groups.* New York, NY: John Wiley, 1990.

8.13 Lie Groups and Symmetric Spaces

* ADAMS, J. FRANK. *Lectures on Lie Groups.* Chicago, IL: University of Chicago Press, 1982.

** CHEVALLEY, CLAUDE. *Theory of Lie Groups.* Princeton, NJ: Princeton University Press, 1946.

DIEUDONNÉ, JEAN. *Special Functions and Linear Representations of Lie Groups.* Providence, RI: American Mathematical Society, 1980.

* HELGASON, SIGURDUR. *Differential Geometry, Lie Groups, and Symmetric Spaces.* New York, NY: Academic Press, 1978.

HELGASON, SIGURDUR. *Groups and Harmonic Analysis.* New York, NY: Academic Press, 1984.

MONTGOMERY, DEANE AND ZIPPIN, LEO. *Topological Transformation Groups.* New York, NY: John Wiley, 1955; Melbourne, FL: Robert E. Krieger, 1974.

SUGIURA, MITSUO. *Unitary Representations and Harmonic Analysis: An Introduction.* New York, NY: Halsted Press, 1975.

* TERRAS, AUDREY. *Harmonic Analysis on Symmetric Spaces and Applications,* 2 Vols. New York, NY: Springer-Verlag, 1985, 1988.

8.14 Nonstandard Analysis

* DAVIS, MARTIN D. *Applied Nonstandard Analysis.* New York, NY: John Wiley, 1977.

HURD, ALBERT E. AND LOEB, PETER A. *An Introduction to Nonstandard Real Analysis.* New York, NY: Academic Press, 1985.

** ROBINSON, ABRAHAM. *Non-standard Analysis.* Amsterdam: North-Holland, 1966.

STROYAN, K.D. AND LUXEMBURG, W.A.J. *Introduction to the Theory of Infinitesimals.* New York, NY: Academic Press, 1976.

8.15 Special Functions

* AKHIEZER, N.I. *Elements of the Theory of Elliptic Functions.* Providence, RI: American Mathematical Society, 1990.

ANDREWS, GEORGE E. *q-Series: Their Development and Applications in Analysis, Number Theory, Combinatorics, Physics, and Computer Algebra.* Providence, RI: American Mathematical Society, 1986.

** ARTIN, EMIL. *The Gamma Function.* New York, NY: Holt, Rinehart and Winston, 1964.

* BAILEY, W.N. *Hypergeometric Series.* New York, NY: Hafner Press, 1972.

** ERDÉLYI, ARTHUR, *et al. Higher Transcendental Functions,* 2 Vols. New York, NY: McGraw-Hill, 1952.

* FINE, NATHAN J. *Basic Hypergeometric Series and Applications.* Providence, RI: American Mathematical Society, 1988.

GASPER, G. AND RAHMAN, M. *Basic Hypergeometric Series.* New York, NY: Cambridge University Press, 1990.

* OLVER, F.W.J. *Asymptotics and Special Functions.* New York, NY: Academic Press, 1974.

RIVLIN, THEODORE J. *Chebyshev Polynomials,* Second Edition. New York, NY: John Wiley, 1974, 1990.

WANG, Z.X. AND GUO, D.R. *Special Functions.* Teaneck, NJ: World Scientific, 1989.

9 Foundations and Mathematical Logic

9.1 Surveys

BETH, EVERT. *The Foundations of Mathematics.* Amsterdam: North-Holland, 1959.

* EVES, HOWARD W. *Foundations and Fundamental Concepts of Mathematics,* Third Edition. Boston, MA: PWS-Kent, 1990.

** MAC LANE, SAUNDERS. *Mathematics, Form and Function.* New York, NY: Springer-Verlag, 1986.

* MOSTOWSKI, ANDRZEJ. *Thirty Years of Foundational Studies.* New York, NY: Barnes and Noble, 1966.

* WILDER, RAYMOND L. *Introduction to the Foundations of Mathematics,* Second Edition. New York, NY: John Wiley, 1965; Melbourne, FL: Robert E. Krieger, 1980.

9.2 Logic

* BARWISE, JON AND ETCHEMENDY, JOHN. *The Liar: An Essay on Truth and Circularity.* New York, NY: Oxford University Press, 1987.

* BOOLE, GEORGE. *An Investigation of the Laws of Thought.* Mineola, NY: Dover, 1951.

COPI, IRVING MARMER. *Symbolic Logic,* Fourth Edition. New York, NY: Macmillan, 1973.

JEFFREY, RICHARD C. *The Logic of Decision,* Second Edition. Chicago, IL: University of Chicago Press, 1983.

QUINE, WILLARD VAN ORMAN. *Methods of Logic,* Third Edition. New York, NY: Holt, Rinehart and Winston, 1972.

SUPPES, PATRICK C. *Introduction to Logic.* New York, NY: Van Nostrand Reinhold, 1957.

VENN, JOHN. *Symbolic Logic,* Second Edition. New York, NY: Chelsea, 1971.

VENN, JOHN. *The Principles of Inductive Logic,* Second Edition. New York, NY: Chelsea, 1973.

9.3 Mathematical Logic

ANDREWS, PETER B. *An Introduction to Mathematical Logic and Type Theory: To Truth Through Proof.* New York, NY: Academic Press, 1986.

* BARWISE, JON AND ETCHEMENDY, JOHN. *The Language of First-Order Logic.* Stanford, CA: Center for Study of Language & Information, 1990.

*** BARWISE, JON, ED. *Handbook of Mathematical Logic.* Amsterdam: North-Holland, 1977.

BELL, J.L. AND MACHOVER, MOSHÉ. *A Course in Mathematical Logic.* Amsterdam: North-Holland, 1977.

* BOOLOS, GEORGE S. AND JEFFREY, RICHARD C. *Computability and Logic,* Third Edition. New York, NY: Cambridge University Press, 1974, 1989.

BOOLOS, GEORGE S. *The Unprovability of Consistency: An Essay in Modal Logic.* New York, NY: Cambridge University Press, 1979.

** CROSSLEY, J.N., *et al. What is Mathematical Logic?* New York, NY: Oxford University Press, 1972.

CURRY, HASKELL B. *Foundations of Mathematical Logic.* Mineola, NY: Dover, 1977.

** DAVIS, MARTIN D., ED. *The Undecidable: Basic Papers on Undecidable Propositions, Unsolvable Problems and Computable Functions.* New York, NY: Raven Press, 1965.

* EBBINGHAUS, H.-D.; FLUM, J.; AND THOMAS, W. *Mathematical Logic.* New York, NY: Springer-Verlag, 1984.

* ENDERTON, HERBERT B. *A Mathematical Introduction to Logic.* New York, NY: Academic Press, 1972.

GOODSTEIN, R.L. *Development of Mathematical Logic.* New York, NY: Springer-Verlag, 1971.

HATCHER, WILLIAM S. *The Logical Foundations of Mathematics.* Philadelphia, PA: Saunders College, 1968; Elmsford, NY: Pergamon Press, 1982.

HILBERT, DAVID. AND ACKERMANN, W. *Principles of Mathematical Logic.* New York, NY: Chelsea, 1950.

*** Kleene, Stephen C. *Introduction to Metamathematics.* Amsterdam: North-Holland, 1974.

* Kleene, Stephen C. *Mathematical Logic.* New York, NY: John Wiley, 1967.

Kneebone, G.T. *Mathematical Logic and the Foundations of Mathematics.* New York, NY: Van Nostrand Reinhold, 1963.

* Mendelson, Elliott. *Introduction to Mathematical Logic,* Third Edition. New York, NY: Van Nostrand Reinhold, 1964; Belmont, CA: Wadsworth, 1987.

* Monk, J. Donald. *Mathematical Logic.* New York, NY: Springer-Verlag, 1976.

** Nagel, Ernest and Newman, James R. *Gödel's Proof.* New York, NY: New York University Press, 1958.

Rosser, J. Barkley. *Logic for Mathematicians,* Second Edition. New York, NY: McGraw-Hill, 1953; New York, NY: Chelsea, 1978.

* Tarski, Alfred. *Logic, Semantics, Metamathematics,* Second Edition. Indianapolis, IN: Hackett, 1983.

9.4 Philosophy of Mathematics

Aspray, William and Kitcher, Philip, eds. *History and Philosophy of Modern Mathematics.* Minneapolis, MN: University of Minnesota Press, 1988.

* Benacerraf, Paul and Putnam, Hilary. *Philosophy of Mathematics: Selected Readings,* Second Edition. Englewood Cliffs, NJ: Prentice Hall, 1964, 1989.

Cohen, Jonathan. *An Introduction to the Philosophy of Induction and Probability.* New York, NY: Oxford University Press, 1989.

Hintikka, Jaakko. *The Philosophy of Mathematics.* New York, NY: Oxford University Press, 1969.

* Kitcher, Philip. *The Nature of Mathematical Knowledge.* New York, NY: Oxford University Press, 1984.

Körner, Stephan. *The Philosophy of Mathematics: An Introductory Essay,* Second Edition. London: Hutchinson, 1960; Mineola, NY: Dover, 1968, 1986.

*** Lakatos, Imre. *Proofs and Refutations: The Logic of Mathematical Discovery.* New York, NY: Cambridge University Press, 1976.

* Quine, Willard Van Orman. *Philosophy of Logic,* Second Edition. Cambridge, MA: Harvard University Press, 1986.

Quine, Willard Van Orman. *Pursuit of Truth.* Cambridge, MA: Harvard University Press, 1990.

Quine, Willard Van Orman. *The Ways of Paradox and Other Essays,* Revised and Enlarged Edition. Cambridge, MA: Harvard University Press, 1976.

** Russell, Bertrand. *Introduction to Mathematical Philosophy.* New York, NY: Macmillan, 1920.

Wang, Hao. *Beyond Analytic Philosophy: Doing Justice to What We Know.* Cambridge, MA: MIT Press, 1986.

* Wittgenstein, Ludwig. *Remarks on the Foundations of Mathematics,* Revised Edition. Cambridge, MA: MIT Press, 1978.

9.5 Set Theory

Baumgartner, James E.; Martin, Donald A.; and Shelah, Saharon, eds. *Axiomatic Set Theory.* Providence, RI: American Mathematical Society, 1984.

Bell, J.L. *Boolean-Valued Models and Independence Proofs in Set Theory.* New York, NY: Clarendon Press, 1977.

Bernays, Paul. *Axiomatic Set Theory,* Mineola, NY: Dover, 1991.

Bourbaki, Nicolas. *Elements of Mathematics: Theory of Sets.* Reading, MA: Addison-Wesley, 1968.

Dales, H.G. and Woodin, W.H. *An Introduction to Independence for Analysts.* New York, NY: Cambridge University Press, 1987.

* Devlin, Keith J. *Fundamentals of Contemporary Set Theory.* New York, NY: Springer-Verlag, 1979.

DEVLIN, KEITH J. *The Axiom of Constructibility: A Guide for the Mathematician.* New York, NY: Springer-Verlag, 1977.

* ENDERTON, HERBERT B. *Elements of Set Theory.* New York, NY: Academic Press, 1977.

** FRAENKEL, ABRAHAM A.; BAR-HILLEL, YEHOSHRA; AND LEVY, AZRIEL. *Foundations of Set Theory,* Second Revised Edition. Atlantic Highlands, NJ: Humanities Press, 1973.

* FRAENKEL, ABRAHAM A. *Abstract Set Theory,* Third Edition. Amsterdam: North Holland, 1953, 1966.

* GÖDEL, KURT. *The Consistency of the Axiom of Choice and of the Generalized Continuum-Hypothesis with the Axioms of Set Theory.* Princeton, NJ: Princeton University Press, 1961.

*** HALMOS, PAUL R. *Naive Set Theory.* New York, NY: Springer-Verlag, 1974.

HRBACEK, KAREL AND JECH, THOMAS. *Introduction to Set Theory,* Second Revised Edition. New York, NY: Marcel Dekker, 1978, 1984.

* JECH, THOMAS. *The Axiom of Choice.* Amsterdam: North-Holland, 1973.

KAMKE, E. *Theory of Sets.* Mineola, NY: Dover, 1950.

** KUNEN, KENNETH. *Set Theory: An Introduction to Independence Proofs,* New York, NY: Elsevier Science, 1983.

* KURATOWSKI, K. AND MOSTOWSKI, ANDRZEJ. *Set Theory with an Introduction to Descriptive Set Theory.* Amsterdam: North-Holland, 1976.

LEVY, AZRIEL. *Basic Set Theory.* New York, NY: Springer-Verlag, 1979.

* MOSCHOVAKIS, YIANNIS N. *Descriptive Set Theory.* Amsterdam: North-Holland, 1980.

QUINE, WILLARD VAN ORMAN. *Set Theory and Its Logic,* Revised Edition. Cambridge, MA: Harvard University Press, 1969.

* ROITMAN, JUDITH. *Introduction to Modern Set Theory.* New York, NY: John Wiley, 1990.

** VILENKIN, N. YA. *Stories About Sets.* New York, NY: Academic Press, 1968.

9.6 Computability

CUTLAND, NIGEL. *Computability: An Introduction to Recursive Function Theory.* New York, NY: Cambridge University Press, 1980.

*** DAVIS, MARTIN D. *Computability and Unsolvability.* New York, NY: McGraw-Hill, 1958; Mineola, NY: Dover, 1982.

ODIFREDDI, P.C. *Classical Recursion Theory.* Amsterdam: North-Holland, 1989.

* POUR-EL, MARIAN B. AND RICHARDS, J. IAN. *Computability in Analysis and Physics.* New York, NY: Springer-Verlag, 1989.

*** ROGERS, HARTLEY, JR. *Theory of Recursive Functions and Effective Computability.* New York, NY: McGraw-Hill, 1967; Cambridge, MA: MIT Press, 1987.

SOARE, ROBERT I. *Recursively Enumerable Sets and Degrees: A Study of Computable Functions and Computably Generated Sets.* New York, NY: Springer-Verlag, 1987.

* USPENSKII, V.A. *Post's Machine.* Moscow: MIR, 1983.

9.7 Model Theory

BALDWIN, JOHN T. *Fundamentals of Stability Theory.* New York, NY: Springer-Verlag, 1988.

* BRIDGE, JANE. *Beginning Model Theory: The Completeness Theorem and Some Consequences.* New York, NY: Clarendon Press, 1977.

*** CHANG, C.C. AND KEISLER, H. JEROME. *Model Theory,* Third Edition. Amsterdam: North-Holland, 1973, 1990.

* HODGES, W. *Building Models by Games.* New York, NY: Cambridge University Press, 1985.

KOPPERMAN, RALPH. *Model Theory and Its Applications.* Boston, MA: Allyn and Bacon, 1972.

LIGHTSTONE, A.H. AND ROBINSON, ABRAHAM. *Nonarchimedean Fields and Asymptotic Expansions.* Amsterdam: North-Holland, 1975.

* LIGHTSTONE, A.H. *Mathematical Logic: An Introduction to Model Theory.* New York, NY: Plenum Press, 1978.

* MORLEY, M.D., ED. *Studies in Model Theory.* Washington, DC: Mathematical Association of America, 1973.

ROBINSON, ABRAHAM. *Complete Theories,* Second Edition. Amsterdam: North-Holland, 1977.

** ROBINSON, ABRAHAM. *Introduction to Model Theory and to the Metamathematics of Algebra.* Amsterdam: North-Holland, 1974.

9.8 Nonclassical Logic

BEESON, MICHAEL J. *Foundations of Constructive Mathematics: Metamathematical Studies.* New York, NY: Springer-Verlag, 1985.

** BISHOP, ERRETT. *Foundations of Constructive Analysis.* New York, NY: McGraw-Hill, 1967.

* BRIDGES, DOUGLAS S. AND RICHMAN, FRED. *Varieties of Constructive Mathematics.* New York, NY: Cambridge University Press, 1987.

DUMMETT, MICHAEL. *Elements of Intuitionism.* New York, NY: Clarendon Press, 1977.

** HEYTING, A. *Intuitionism: An Introduction,* Third Revised Edition. Amsterdam: North-Holland, 1976.

** TROELSTRA, A.S. AND VAN DALEN, DIRK. *Constructivism in Mathematics: An Introduction.* Amsterdam: North-Holland, 1988.

VAN DALEN, DIRK, ED. *Brouwer's Cambridge Lectures on Intuitionism.* New York, NY: Cambridge University Press, 1981.

9.9 Special Topics

FREGE, GOTTLOB. *On the Foundations of Geometry and Formal Theories of Arithmetic.* New Haven, CT: Yale University Press, 1971.

GOLDBLATT, ROBERT. *Topoi: The Categorical Analysis of Logic.* Amsterdam: North-Holland, 1979.

* GONSHOR, HARRY. *An Introduction to the Theory of Surreal Numbers.* New York, NY: Cambridge University Press, 1986.

HENKIN, LEON; MONK, J. DONALD; AND TARSKI, ALFRED. *Cylindric Algebras,* 2 Vols. Amsterdam: North-Holland, 1971.

HINDLEY, J. ROGER AND SELDIN, JONATHAN P. *Introduction to Combinators and λ-Calculus.* New York, NY: Cambridge University Press, 1986.

* TAKEUTI, GAISI. *Proof Theory.* Amsterdam: North-Holland, 1975.

* TARSKI, ALFRED. *Ordinal Algebras.* Amsterdam: North-Holland, 1956.

10 Discrete Mathematics

10.1 Discrete Mathematics

* ALBERTSON, MICHAEL O. AND HUTCHINSON, JOAN P. *Discrete Mathematics with Algorithms.* New York, NY: John Wiley, 1988.

ALTHOEN, STEVEN C. AND BUMCROT, ROBERT J. *Introduction to Discrete Mathematics.* Boston, MA: PWS-Kent, 1988.

* BIGGS, NORMAN L. *Discrete Mathematics,* Revised Edition. New York, NY: Clarendon Press, 1985, 1989.

* BOGART, KENNETH P. *Discrete Mathematics.* Lexington, MA: D.C. Heath, 1988.

DOERR, ALAN AND LEVASSEUR, KENNETH. *Applied Discrete Structures for Computer Science.* Chicago, IL: Science Research Association, 1985.

DOSSEY, JOHN A., *et al.* *Discrete Mathematics.* Glenview, IL: Scott Foresman, 1986.

* EPP, SUSANNA S. *Discrete Mathematics with Applications.* Belmont, CA: Wadsworth, 1990.

FINKBEINER, DANIEL T. AND LINDSTROM, WENDELL D. *A Primer of Discrete Mathematics.* New York, NY: W.H. Freeman, 1987.

* GERSTEIN, LARRY J. *Discrete Mathematics and Algebraic Structures.* New York, NY: W.H. Freeman, 1987.

GRIMALDI, RALPH P. *Discrete and Combinatorial Mathematics: An Applied Introduction,* Second Edition. Reading, MA: Addison-Wesley, 1989.

* JOHNSONBAUGH, RICHARD. *Discrete Mathematics,* Second Edition. New York, NY: Macmillan, 1990.

** MAURER, STEPHEN B. AND RALSTON, ANTHONY. *Discrete Algorithmic Mathematics.* Reading, MA: Addison-Wesley, 1991.

MCELIECE, ROBERT J.; ASH, ROBERT B.; AND ASH, CAROL. *Introduction to Discrete Mathematics.* Cambridge, MA: Random House, 1989.

NICODEMI, OLYMPIA. *Discrete Mathematics: A Bridge to Computer Science and Advanced Mathematics.* St. Paul, MN: West, 1987.

POLIMENI, ALBERT D. AND STRAIGHT, H. JOSEPH. *Foundations of Discrete Mathematics,* Second Edition. Pacific Grove, CA: Brooks/Cole, 1985, 1990.

PRATHER, RONALD. *Elements of Discrete Mathematics.* Boston, MA: Houghton Mifflin, 1986.

* ROMAN, STEVEN. *An Introduction to Discrete Mathematics,* Second Edition. Philadelphia, PA: Saunders College, 1986; San Diego, CA: Harcourt Brace Jovanovich, 1989.

ROSEN, KENNETH H. *Discrete Mathematics and its Applications.* Cambridge, MA: Random House, 1988.

** ROSS, KENNETH A. AND WRIGHT, CHARLES R.B. *Discrete Mathematics,* Second Edition. Englewood Cliffs, NJ: Prentice Hall, 1985, 1988.

10.2 Finite Mathematics

* ANTON, HOWARD; KOLMAN, BERNARD; AND AVERBACH, BONNIE. *Applied Finite Mathematics,* Fourth Edition. New York, NY: Academic Press, 1982, 1988.

* BITTINGER, MARVIN L. AND CROWN, J. CONRAD. *Finite Mathematics.* Reading, MA: Addison-Wesley, 1989.

COZZENS, MARGARET B. AND PORTER, RICHARD D. *Mathematics and Its Applications to Management, Life, and Social Sciences With Finite and Discrete Mathematics.* Lexington, MA: D.C. Heath, 1987.

* DAVIS, MORTON D. *The Art of Decision-Making.* New York, NY: Springer-Verlag, 1986.

FARLOW, STANLEY J. AND HAGGARD, GARY M. *Finite Mathematics and Its Applications.* Cambridge, MA: Random House, 1988.

** GOLDSTEIN, LARRY J.; SCHNEIDER, DAVID I.; AND SIEGEL, MARTHA J. *Finite Mathematics and Its Applications,* Fourth Edition. Englewood Cliffs, NJ: Prentice Hall, 1984, 1991.

HOENIG, ALAN. *Applied Finite Mathematics.* New York, NY: McGraw-Hill, 1990.

*** KEMENY, JOHN G.; SNELL, J. LAURIE; AND THOMPSON, GERALD L. *Introduction to Finite Mathematics,* Third Edition. Englewood Cliffs, NJ: Prentice Hall, 1974.

* MAKI, DANIEL P. AND THOMPSON, MAYNARD. *Finite Mathematics,* Third Edition. New York, NY: McGraw-Hill, 1978, 1989.

MALKEVITCH, JOSEPH AND MEYER, WALTER. *Graphs, Models, and Finite Mathematics.* Englewood Cliffs, NJ: Prentice Hall, 1974.

SMITH, KARL J. *Finite Mathematics,* Second Edition. Pacific Grove, CA: Brooks/Cole, 1988.

SPENCE, LAWRENCE E.; VANDEN EYNDEN, CHARLES; AND GALLIN, DANIEL. *Finite Mathematics.* Glenview, IL: Scott Foresman, 1990.

* TAN, S.T. *Applied Finite Mathematics,* Third Edition. Boston, MA: Prindle, Weber and Schmidt, 1983, 1990.

10.3 Introductory Combinatorics

* ANDERSON, IAN. *A First Course in Combinatorial Mathematics.* New York, NY: Oxford University Press, 1974.

BERGE, CLAUDE. *Principles of Combinatorics.* New York, NY: Academic Press, 1971.

BOGART, KENNETH P. *Introductory Combinatorics,* Second Edition. Brooklyn, NY: Pitman, 1983; San Diego, CA: Harcourt Brace Jovanovich, 1990.

BOSE, R.C. AND MANVEL, B. *Introduction to Combinatorial Theory.* New York, NY: John Wiley, 1984.

** BRUALDI, RICHARD A. *Introductory Combinatorics,* Second Edition. Amsterdam: North-Holland, 1977, 1991.

* COHEN, DANIEL I.A. *Basic Techniques of Combinatorial Theory.* New York, NY: John Wiley, 1978.

*** GRAHAM, RONALD L.; KNUTH, DONALD E.; AND PATASHNIK, OREN. *Concrete Mathematics: A Foundation for Computer Science.* Reading, MA: Addison-Wesley, 1989.

* HILLMAN, ABRAHAM P.; ALEXANDERSON, GERALD L.; AND GRASSL, RICHARD M. *Discrete and Combinatorial Mathematics.* San Francisco, CA: Dellen, 1987.

JACKSON, BRAD AND THORO, DMITRI. *Applied Combinatorics with Problem Solving.* Reading, MA: Addison-Wesley, 1990.

* LIU, C.L. *Introduction to Applied Combinatorial Mathematics.* New York, NY: McGraw-Hill, 1968.

*** NIVEN, IVAN M. *Mathematics of Choice or How to Count Without Counting.* Washington, DC: Mathematical Association of America, 1975.

PAGE, E.S. AND WILSON, L.B. *An Introduction to Computational Combinatorics.* New York, NY: Cambridge University Press, 1979.

** ROBERTS, FRED S. *Applied Combinatorics.* Englewood Cliffs, NJ: Prentice Hall, 1984.

** STANTON, DENNIS AND WHITE, DENNIS. *Constructive Combinatorics.* New York, NY: Springer-Verlag, 1986.

** TUCKER, ALAN. *Applied Combinatorics.* New York, NY: John Wiley, 1980, 1984.

10.4 Advanced Combinatorics

AIGNER, MARTIN. *Combinatorial Search.* New York, NY: John Wiley, 1988.

AIGNER, MARTIN. *Combinatorial Theory.* New York, NY: Springer-Verlag, 1979.

* ANDERSON, IAN. *Combinatorial Designs: Construction Methods.* New York, NY: Ellis Horwood, 1990.

** ANDERSON, IAN. *Combinatorics of Finite Sets.* New York, NY: Oxford University Press, 1987.

BOLLOBÁS, BÉLA. *Combinatorics: Set Systems, Hypergraphs, Families of Vectors, and Combinatorial Probability.* New York, NY: Cambridge University Press, 1986.

BRUALDI, RICHARD A. AND RYSER, H.J. *Combinatorial Matrix Theory.* New York, NY: Cambridge University Press, 1991.

** CONWAY, JOHN HORTON. *On Numbers and Games.* New York, NY: Academic Press, 1976.

* DENES, J. AND KEEDWELL, A.D. *Latin Squares and Their Applications.* New York, NY: Academic Press, 1974.

* ERDŐS, P. AND SPENCER, JOEL H. *Probability Methods in Combinatorics.* New York, NY: Academic Press, 1974.

GOULDEN, I.P. AND JACKSON, D.M. *Combinatorial Enumeration.* New York, NY: John Wiley, 1983.

** GRAHAM, RONALD L.; ROTHSCHILD, BRUCE L.; AND SPENCER, JOEL H. *Ramsey Theory,* Second Edition. New York, NY: John Wiley, 1980, 1990.

GREENE, DANIEL H. AND KNUTH, DONALD E. *Mathematics for the Analysis of Algorithms,* Third Edition. New York, NY: Birkhäuser, 1981, 1990.

GUSFIELD, DAN AND IRVING, ROBERT W. *The Stable Marriage Problem: Structure and Algorithms.* Cambridge, MA: MIT Press, 1989.

* HALL, MARSHALL, JR. *Combinatorial Theory,* Second Edition. New York, NY: Blaisdell, 1967; New York, NY: John Wiley, 1986.

LOVÁSZ, LÁSZLÓ. *An Algorithmic Theory of Numbers, Graphs, and Convexity.* Philadelphia, PA: Society for Industrial and Applied Mathematics, 1986.

*** LOVÁSZ, LÁSZLÓ. *Combinatorial Problems and Exercises.* Amsterdam: North-Holland, 1979.

PÓLYA, GEORGE AND READ, RONALD C. *Combinatorial Enumeration of Groups, Graphs, and Chemical Compounds.* New York, NY: Springer-Verlag, 1987.

PÓLYA, GEORGE; TARJAN, ROBERT E.; AND WOODS, DONALD R. *Notes on Introductory Combinatorics.* New York, NY: Birkhäuser, 1983.

RAY-CHAUDHURI, D.K., ED. *Relations Between Combinatorics and Other Parts of Mathematics.* Providence, RI: American Mathematical Society, 1979.

RIORDAN, JOHN. *An Introduction to Combinatorial Analysis.* New York, NY: John Wiley, 1958; Princeton, NJ: Princeton University Press, 1978.

RIORDAN, JOHN. *Combinatorial Identities.* New York, NY: John Wiley, 1968.

ROTA, GIAN-CARLO, ED. *Studies in Combinatorics.* Washington, DC: Mathematical Association of America, 1978.

*** RYSER, H.J. *Combinatorial Mathematics.* Washington, DC: Mathematical Association of America, 1963.

** STANLEY, RICHARD P. *Enumerative Combinatorics,* Belmont, CA: Wadsworth, 1986.

* STREET, ANNE P. AND STREET, DEBORAH J. *Combinatorics of Experimental Design.* New York, NY: Clarendon Press, 1987.

TOMESCU, IOAN. *Problems in Combinatorics and Graph Theory.* New York, NY: John Wiley, 1985.

* VILENKIN, N. YA. *Combinatorics.* New York, NY: Academic Press, 1971.

WALLIS, W.D. *Combinatorial Designs.* New York, NY: Marcel Dekker, 1988.

10.5 Graph Theory

BARNETTE, DAVID. *Map Coloring, Polyhedra, and the Four-Color Problem.* Washington, DC: Mathematical Association of America, 1983.

BEINEKE, LOWELL W. AND WILSON, ROBIN J., EDS. *Selected Topics in Graph Theory,* 3 Vols. New York, NY: Academic Press, 1978–88.

* BERGE, CLAUDE. *Graphs,* Second Revised Edition. New York, NY: Elsevier Science, 1985.

BERGE, CLAUDE. *Hypergraphs: Combinatorics of Finite Sets.* Amsterdam: North-Holland, 1989.

** BIGGS, NORMAN L.; LLOYD, E. KEITH; AND WILSON, ROBIN J. *Graph Theory, 1736–1936.* New York, NY: Clarendon Press, 1976; New York, NY: Oxford University Press, 1986.

BIGGS, NORMAN L. *Algebraic Graph Theory.* New York, NY: Cambridge University Press, 1974.

* BOLLOBÁS, BÉLA. *Graph Theory: An Introductory Course.* New York, NY: Springer-Verlag, 1979.

BOLLOBÁS, BÉLA. *Random Graphs.* New York, NY: Academic Press, 1985.

*** BONDY, J. ADRIAN AND MURTY, U.S.R. *Graph Theory with Applications.* New York, NY: American Elsevier, 1976.

BUCKLEY, FRED AND HARARY, FRANK. *Distance in Graphs.* Reading, MA: Addison-Wesley, 1990.

* CAPOBIANCO, M. AND MOLLUZZO, J. *Examples and Counterexamples in Graph Theory.* Amsterdam: North-Holland, 1978.

*** CHARTRAND, GARY AND LESNIAK, LINDA. *Graphs & Digraphs,* Second Edition. Boston, MA: Prindle, Weber and Schmidt, 1979; Belmont, CA: Wadsworth, 1986.

* CHARTRAND, GARY. *Introductory Graph Theory.* (Former title: *Graphs as Mathematical Models.*) Boston, MA: Prindle, Weber and Schmidt, 1977; Mineola, NY: Dover, 1985.

* FULKERSON, D.R., ED. *Studies in Graph Theory.* Washington, DC: Mathematical Association of America, 1975.

GIBBONS, ALAN. *Algorithmic Graph Theory.* New York, NY: Cambridge University Press, 1985.

** GOULD, RONALD. *Graph Theory.* Redwood City, CA: Benjamin Cummings, 1988.

GROSS, JONATHAN L. AND TUCKER, THOMAS W. *Topological Graph Theory.* New York, NY: John Wiley, 1987.

HARARY, FRANK. *Graph Theory.* Reading, MA: Addison-Wesley, 1969.

HARTSFIELD, NORA AND RINGEL, GERHARD. *Pearls in Graph Theory: A Comprehensive Introduction.* New York, NY: Academic Press, 1990.

KÖNIG, DÉNES. *Theory of Finite and Infinite Graphs.* New York, NY: Birkhäuser, 1990.

* LOVÁSZ, LÁSZLÓ AND PLUMMER, M. *Matching Theory.* Amsterdam: North-Holland, 1986.

* ORE, OYSTEIN. *Graphs and Their Uses.* Washington, DC: Mathematical Association of America, 1963, 1990.

* PALMER, EDGAR M. *Graphical Evolution: An Introduction to the Theory of Random Graphs.* New York, NY: John Wiley, 1985.

* SAATY, THOMAS L. AND KAINEN, PAUL C. *The Four-Color Problem: Assaults and Conquest.* New York, NY: McGraw-Hill, 1977.

STEINBACH, PETER. *Field Guide to Simple Graphs.* Albuquerque, NM: Design Lab, 1990.

TRUDEAU, RICHARD J. *Dots and Lines.* Kent, OH: Kent State University Press, 1976.

TUTTE, W.T. *Graph Theory.* Reading, MA: Addison-Wesley, 1984.

WHITE, ARTHUR T. *Graphs, Groups and Surfaces.* New York, NY: Elsevier Science, 1984.

** WILSON, ROBIN J. AND WATKINS, J. *Graphs: An Introductory Approach.* New York, NY: John Wiley, 1990.

10.6 Coding Theory

BEKER, HENRY AND PIPER, FRED. *Cipher Systems: The Protection of Communications.* New York, NY: John Wiley, 1982.

BLAHUT, RICHARD E. *Theory and Practice of Error Control Codes.* Reading, MA: Addison-Wesley, 1983.

BLAKE, IAN F. AND MULLIN, RONALD C. *An Introduction to Algebraic and Combinatorial Coding Theory.* New York, NY: Academic Press, 1976.

BRASSARD, GILLES. *Modern Cryptology: A Tutorial.* New York, NY: Springer-Verlag, 1988.

CAMERON, P.J. AND VAN LINT, J.H. *Graphs, Codes and Designs.* New York, NY: Cambridge University Press, 1980.

GOPPA, V.D. *Geometry and Codes.* Norwell, MA: Kluwer Academic, 1988.

** HAMMING, RICHARD W. *Coding and Information Theory,* Second Edition. Englewood Cliffs, NJ: Prentice Hall, 1986.

* HILL, RAYMOND. *A First Course in Coding Theory.* New York, NY: Clarendon Press, 1986.

KONHEIM, ALAN G. *Cryptography: A Primer.* New York, NY: John Wiley, 1981.

** MACWILLIAMS, F.J. AND SLOANE, N.J.A. *The Theory of Error-Correcting Codes.* Amsterdam: North-Holland, 1977.

* PETERSON, W. WESLEY AND WELDON, E.J., JR. *Error-Correcting Codes,* Second Edition. Cambridge, MA: MIT Press, 1961, 1972.

*** PLESS, VERA. *Introduction to the Theory of Error-Correcting Codes,* Second Edition. New York, NY: John Wiley, 1982, 1989.

SLOANE, N.J.A. *A Short Course on Error Correcting Codes.* New York, NY: Springer-Verlag, 1975.

*** THOMPSON, THOMAS M. *From Error-Correcting Codes Through Sphere Packings to Simple Groups.* Washington, DC: Mathematical Association of America, 1983.

* WELSH, DOMINIC. *Codes and Cryptography.* New York, NY: Clarendon Press, 1988.

10.7 Special Topics

BETH, THOMAS; JUNGNICKEL, DIETER; AND LENZ, HANFRIED. *Design Theory.* Mannheim: Bibliographisches Institut, 1985; New York, NY: Cambridge University Press, 1986.

DAVEY, B.A. AND PRIESTLEY, H.A. *Introduction to Lattices and Order.* New York, NY: Cambridge University Press, 1990.

** GOLDBERG, SAMUEL I. *Introduction to Difference Equations.* New York, NY: John Wiley, 1958; Mineola, NY: Dover, 1986.

GRÄTZER, GEORGE. *General Lattice Theory.* New York, NY: Academic Press, 1978.

MICKENS, RONALD E. *Difference Equations.* New York, NY: Van Nostrand Reinhold, 1987.

* WELSH, D.J.A. *Matroid Theory.* New York, NY: Academic Press, 1976.

WHITE, NEIL, ED. *Theory of Matroids.* New York, NY: Cambridge University Press, 1986.

** WILF, HERBERT S. *Generating Functionology.* New York, NY: Academic Press, 1990.

11 Number Theory

11.1 Introductory Texts

* ADAMS, WILLIAM W. AND GOLDSTEIN, LARRY J. *Introduction to Number Theory.* Englewood Cliffs, NJ: Prentice Hall, 1976.

ALLENBY, R.B.J.T. AND REDFERN, E.J. *Introduction to Number Theory with Computing.* New York, NY: Edward Arnold, 1989.

** BAKER, ALAN. *A Concise Introduction to the Theory of Numbers.* New York, NY: Cambridge University Press, 1984.

BURN, R.P. *A Pathway Into Number Theory.* New York, NY: Cambridge University Press, 1982.

BURTON, DAVID M. *Elementary Number Theory,* Second Edition. Needham Heights, MA: Allyn and Bacon, 1976; Dubuque, IA: William C. Brown, 1989.

*** DAVENPORT, HAROLD. *The Higher Arithmetic: An Introduction to the Theory of Numbers,* Fifth Edition. Atlantic Highlands, NJ: Humanities Press, 1968; New York, NY: Cambridge University Press, 1982.

DICKSON, LEONARD E. *Introduction to the Theory of Numbers.* Mineola, NY: Dover, 1959.

DUDLEY, UNDERWOOD. *Elementary Number Theory,* Second Edition. New York, NY: W.H. Freeman, 1978.

FLATH, DANIEL E. *Introduction to Number Theory.* New York, NY: John Wiley, 1989.

MCCOY, NEAL H. *The Theory of Numbers.* New York, NY: Macmillan, 1965; New York, NY: Chelsea, 1985.

* NAGELL, TRYGVE. *Introduction to Number Theory.* New York, NY: Chelsea, 1964.

*** NIVEN, IVAN M.; ZUCKERMAN, HERBERT S.; AND MONTGOMERY, HUGH L. *An Introduction to the Theory of Numbers,* Fifth Edition. New York, NY: John Wiley, 1966, 1991.

ROSE, H.E. *A Course in Number Theory.* New York, NY: Clarendon Press, 1988.

ROSEN, KENNETH H. *Elementary Number Theory and Its Applications,* Second Edition. Reading, MA: Addison-Wesley, 1984, 1988.

SHAPIRO, HAROLD N. *Introduction to the Theory of Numbers.* New York, NY: John Wiley, 1983.

* STARK, HAROLD M. *An Introduction to Number Theory.* Chicago, IL: Rand McNally, 1970; Cambridge, MA: MIT Press, 1978.

** USPENSKY, J.V. AND HEASLET, M.A. *Elementary Number Theory.* New York, NY: McGraw-Hill, 1939.

VINOGRADOV, I.M. *Elements of Number Theory.* Mineola, NY: Dover, 1954.

*** WEIL, ANDRÉ. *Number Theory for Beginners.* New York, NY: Springer-Verlag, 1979.

11.2 Expositions

BESKIN, N.M. *Fascinating Fractions.* Moscow: MIR, 1986.

* KHINCHIN, A. YA. *Three Pearls of Number Theory.* Baltimore, MD: Graylock Press, 1952.

** NIVEN, IVAN M. *Irrational Numbers.* Washington, DC: Mathematical Association of America, 1959.

*** NIVEN, IVAN M. *Numbers: Rational and Irrational.* Washington, DC: Mathematical Association of America, 1961.

OGILVY, C. STANLEY AND ANDERSON, JOHN T. *Excursions in Number Theory.* New York, NY: Oxford University Press, 1966.

ORE, OYSTEIN. *Invitation to Number Theory.* Washington, DC: Mathematical Association of America, 1967, 1975.

** RADEMACHER, HANS. *Lectures on Elementary Number Theory.* Melbourne, FL: Robert E. Krieger, 1977.

* SCHARLAU, WINFRIED AND OPOLKA, HANS. *From Fermat to Minkowski: Lectures on the Theory of Numbers and Its Historical Development.* New York, NY: Springer-Verlag, 1985.

** SCHROEDER, MANFRED R. *Number Theory in Science and Communication with Applications in Cryptography, Physics, Digital Information, Computing, and Self-Similarity*, Second Enlarged Edition. New York, NY: Springer-Verlag, 1984, 1986.

*** WEIL, ANDRÉ. *Number Theory: An Approach Through History From Hammurapi to Legendre.* New York, NY: Birkhäuser, 1984.

11.3 Elementary Monographs

* ANDREWS, GEORGE E. *Number Theory.* New Delhi: Hindustan Publishing, 1985.

* DAVIS, PHILIP J. *The Lore of Large Numbers.* Washington, DC: Mathematical Association of America, 1975.

DICKSON, LEONARD E. *Modern Elementary Theory of Numbers.* Chicago, IL: University of Chicago Press, 1939.

* ERDŐS, P. AND GRAHAM, RONALD L. *Old and New Problems and Results in Combinatorial Number Theory.* Geneva: L'Enseignement Mathématique, 1980.

* GROSSWALD, EMIL. *Topics from the Theory of Numbers*, Second Edition. New York, NY: Birkhäuser, 1966, 1984.

*** GUY, RICHARD K. *Unsolved Problems in Number Theory.* New York, NY: Springer-Verlag, 1981.

*** HARDY, G.H. AND WRIGHT, E.M. *Introduction to the Theory of Numbers*, Fifth Edition. New York, NY: Oxford University Press, 1960, 1979.

** HUA, LOO-KENG. *Introduction to Number Theory.* New York, NY: Springer-Verlag, 1982.

HURWITZ, ADOLF AND KRITIKOS, NIKOLAOS. *Lectures on Number Theory.* New York, NY: Springer-Verlag, 1986.

* KHINCHIN, A. YA. *Continued Fractions.* Chicago, IL: University of Chicago Press, 1964.

KOBLITZ, NEAL. *A Course in Number Theory and Cryptography.* New York, NY: Springer-Verlag, 1987.

** LANDAU, EDMUND G.H. *Elementary Number Theory*, Second Edition. New York, NY: Chelsea, 1958, 1966.

* LEVEQUE, WILLIAM J. *Topics in Number Theory*, 2 Vols. Reading, MA: Addison-Wesley, 1956.

MATHEWS, G.B. *Theory of Numbers.* New York, NY: Chelsea, 1961.

MCCARTHY, PAUL J. *Introduction to Arithmetical Functions.* New York, NY: Springer-Verlag, 1986.

OLDS, C.D. *Continued Fractions.* Washington, DC: Mathematical Association of America, 1963.

* ORE, OYSTEIN. *Number Theory and Its History.* Mineola, NY: Dover, 1988.

** SHANKS, DANIEL. *Solved and Unsolved Problems in Number Theory*, Third Edition. East Lansing, MI: Spartan Press, 1962; New York, NY: Chelsea, 1978, 1985.

* SIERPIŃSKI, W. *250 Problems in Elementary Number Theory.* New York, NY: American Elsevier, 1970.

* SIERPIŃSKI, W. *Elementary Theory of Numbers.* Amsterdam: North-Holland, 1988.

SIERPIŃSKI, W. *Theory of Numbers.* New York, NY: International Publication Service, 1973.

VOROB'EV, N.N. *Fibonacci Numbers.* New York, NY: Blaisdell, 1961.

11.4 Primes and Factors

* BRESSOUD, DAVID M. *Factorization and Primality Testing.* New York, NY: Springer-Verlag, 1989.

BRILLHART, JOHN, *et al.* *Factorizations of* b^n+1, $b = 2,3,5,6,7,10,11,12$ *up to High Powers.* Providence, RI: American Mathematical Society, 1983.

* ELLISON, WILLIAM J. AND ELLISON, FERN. *Prime Numbers.* New York, NY: John Wiley, 1985.

* HALBERSTAM, H. AND RICHERT, H.-E. *Sieve Methods.* New York, NY: Academic Press, 1974.

** POMERANCE, CARL. *Lecture Notes on Primality Testing and Factoring: A Short Course at Kent State University.* Washington, DC: Mathematical Association of America, 1984.

** RIBENBOIM, PAULO. *The Book of Prime Number Records*, Second Edition. New York, NY: Springer-Verlag, 1988, 1989.

* RIBENBOIM, PAULO. *The Little Book of Big Primes.* New York, NY: Springer-Verlag, 1991.

RIESEL, HANS. *Prime Numbers and Computer Methods for Factorization.* New York, NY: Birkhäuser, 1985.

11.5 Algebraic Number Theory

ARTIN, EMIL AND TATE, JOHN. *Class Field Theory.* Reading, MA: Addison-Wesley, 1990.

* BAKER, ALAN. *Transcendental Number Theory.* New York, NY: Cambridge University Press, 1975.

*** BOREVICH, Z.I. AND SHAFAREVICH, IGOR R. *Number Theory.* New York, NY: Academic Press, 1966.

BUELL, DUNCAN A. *Binary Quadratic Forms: Classical Theory and Modern Computations.* New York, NY: Springer-Verlag, 1989.

CASSELS, J.W.S. *Rational Quadratic Forms.* New York, NY: Academic Press, 1978.

COHN, HARVEY. *A Classical Invitation to Algebraic Numbers and Class Fields.* New York, NY: Springer-Verlag, 1978.

COX, DAVID A. *Primes of the Form x^2+ny^2: Fermat, Class Field Theory, and Complex Multiplication.* New York, NY: John Wiley, 1989.

DICKSON, LEONARD E., *et al. Algebraic Numbers.* New York, NY: Chelsea, 1967.

DICKSON, LEONARD E. *Studies in the Theory of Numbers.* New York, NY: Chelsea, 1957.

EDWARDS, HAROLD M. *Divisor Theory.* New York, NY: Birkhäuser, 1990.

*** EDWARDS, HAROLD M. *Fermat's Last Theorem: A Genetic Introduction to Algebraic Number Theory.* New York, NY: Springer-Verlag, 1977.

* GELFOND, A.O. *Transcendental and Algebraic Numbers.* Mineola, NY: Dover, 1960.

* HASSE, HELMUT. *Number Theory.* New York, NY: Springer-Verlag, 1980.

** HECKE, ERICH. *Lectures on the Theory of Algebraic Numbers.* New York, NY: Springer-Verlag, 1981.

** IRELAND, KENNETH AND ROSEN, MICHAEL I. *A Classical Introduction to Modern Number Theory.* Belmont, CA: Bogden and Quigley, 1972; New York, NY: Springer-Verlag, 1982.

JANUSZ, GERALD J. *Algebraic Number Fields.* New York, NY: Academic Press, 1973.

** LANG, SERGE. *Algebraic Number Theory.* New York, NY: Springer-Verlag, 1986.

MARCUS, DANIEL A. *Number Fields.* New York, NY: Springer-Verlag, 1977.

O'MEARA, O.T. *Introduction to Quadratic Forms.* New York, NY: Springer-Verlag, 1963.

POHST, M. AND ZASSENHAUS, H. *Algorithmic Algebraic Number Theory.* New York, NY: Cambridge University Press, 1989.

** POLLARD, HARRY AND DIAMOND, H.G. *The Theory of Algebraic Numbers,* Second Edition. Washington, DC: Mathematical Association of America, 1976.

* RIBENBOIM, PAULO. *13 Lectures on Fermat's Last Theorem.* New York, NY: Springer-Verlag, 1979.

SAMUEL, PIERRE. *Algebraic Theory of Numbers.* Boston, MA: Houghton Mifflin, 1970.

* SERRE, JEAN-PIERRE. *A Course in Arithmetic.* New York, NY: Springer-Verlag, 1973.

STEWART, IAN AND TALL, DAVID. *Algebraic Number Theory,* Second Edition. New York, NY: Chapman and Hall, 1979, 1987.

* WASHINGTON, LAWRENCE C. *Introduction to Cyclotomic Fields.* New York, NY: Springer-Verlag, 1982.

WEIL, ANDRÉ. *Basic Number Theory.* New York, NY: Springer-Verlag, 1973.

11.6 Analytic Number Theory

*** APOSTOL, TOM M. *Introduction to Analytic Number Theory.* New York, NY: Springer-Verlag, 1976.

* AYOUB, R. *An Introduction to the Analytic Theory of Numbers.* Providence, RI: American Mathematical Society, 1963.

BELLMAN, RICHARD E. *Analytic Number Theory: An Introduction.* Redwood City, CA: Benjamin Cummings, 1980.

* BORWEIN, JONATHAN M. AND BORWEIN, PETER B. *Pi and the AGM: A Study in Analytic Number Theory and Computational Complexity.* New York, NY: John Wiley, 1987.

CHANDRASEKHARAN, KOMARAVOLU. *Introduction to Analytic Number Theory.* New York, NY: Springer-Verlag, 1968.

** DAVENPORT, HAROLD. *Multiplicative Number Theory,* Second Edition. Chicago, IL: Rand McNally, 1967; New York, NY: Springer-Verlag, 1980.

* EDWARDS, HAROLD M. *Riemann's Zeta Function.* New York, NY: Academic Press, 1974.

* GROSSWALD, EMIL. *Representations of Integers as Sums of Squares.* New York, NY: Springer-Verlag, 1985.

* INGHAM, A.E. *The Distribution of Prime Numbers.* New York, NY: Hafner Press, 1971.

IVIC, A. *The Riemann Zeta-Function.* New York, NY: John Wiley, 1985.

RADEMACHER, HANS. *Topics in Analytic Number Theory.* New York, NY: Springer-Verlag, 1973.

** TITCHMARSH, EDWARD C. *The Theory of the Riemann Zeta-Function,* Second Edition. New York, NY: Clarendon Press, 1986.

11.7 Modular Forms

* APOSTOL, TOM M. *Modular Functions and Dirichlet Series in Number Theory,* Second Edition. New York, NY: Springer-Verlag, 1976, 1990.

CHANDRASEKHARAN, KOMARAVOLU. *Elliptic Functions.* New York, NY: Springer-Verlag, 1985.

HUSEMÖLLER, DALE. *Elliptic Curves.* New York, NY: Springer-Verlag, 1986.

* KOBLITZ, NEAL. *Introduction to Elliptic Curves and Modular Forms.* New York, NY: Springer-Verlag, 1984.

LANG, SERGE. *Elliptic Functions,* Second Edition. New York, NY: Springer-Verlag, 1987.

SILVERMAN, JOSEPH H. *The Arithmetic of Elliptic Curves.* New York, NY: Springer-Verlag, 1986.

WEIL, ANDRÉ. *Elliptic Functions According to Eisenstein and Kronecker.* New York, NY: Springer-Verlag, 1976.

11.8 P-adic Fields

BACHMAN, GEORGE. *Introduction to p-adic Numbers and Valuation Theory.* New York, NY: Academic Press, 1964.

KOBLITZ, NEAL. *p-adic Analysis: A Short Course on Recent Work.* New York, NY: Cambridge University Press, 1980.

KOBLITZ, NEAL. *p-adic Numbers, p-adic Analysis, and Zeta-Functions,* Second Edition. New York, NY: Springer-Verlag, 1984.

SCHIKHOF, W.H. *Ultrametric Calculus: An Introduction to p-adic Analysis.* New York, NY: Cambridge University Press, 1984.

SERRE, JEAN-PIERRE. *Local Fields.* New York, NY: Springer-Verlag, 1979

11.9 Special Topics

** ANDREWS, GEORGE E. *The Theory of Partitions,* Second Edition. Reading, MA: Addison-Wesley, 1976; New York, NY: Cambridge University Press, 1985.

CARMICHAEL, ROBERT D. *The Theory of Numbers and Diophantine Analysis.* Mineola, NY: Dover, 1959.

CASSELS, J.W.S. *An Introduction to the Geometry of Numbers.* New York, NY: Springer-Verlag, 1959, 1971.

DYNKIN, E.B. AND USPENSKII, V.A. *Problems in the Theory of Numbers.* Lexington, MA: D.C. Heath, 1963.

EBBINGHAUS, H.-D., *et al. Numbers.* New York, NY: Springer-Verlag, 1990.

ELLIOTT, P.D.T.A. *Probabilistic Number Theory,* 2 Vols. New York, NY: Springer-Verlag, 1979, 1980.

MORDELL, L.J. *Diophantine Equations.* New York, NY: Academic Press, 1969.

PARENT, D.P. *Exercises in Number Theory.* New York, NY: Springer-Verlag, 1984.

SIERPIŃSKI, W. *A Selection of Problems in the Theory of Numbers.* Elmsford, NY: Pergamon Press, 1964.

WANG, YUAN. *Goldbach Conjecture.* Teaneck, NJ: World Scientific, 1984.

12 Linear Algebra

12.1 Elementary

** Anton, Howard. *Elementary Linear Algebra*, Sixth Edition. New York, NY: John Wiley, 1973, 1991.

** Banchoff, Thomas F. and Wermer, John. *Linear Algebra Through Geometry.* New York, NY: Springer-Verlag, 1983.

Bloom, David M. *Linear Algebra and Geometry.* New York, NY: Cambridge University Press, 1979.

* Curtis, Charles W. *Linear Algebra: An Introductory Approach*, Fourth Edition. Boston, MA: Allyn and Bacon, 1974; New York, NY: Springer-Verlag, 1984.

Damiano, David B. and Little, John B. *A Course in Linear Algebra.* San Diego, CA: Harcourt Brace Jovanovich, 1988.

Fraleigh, John B. and Beauregard, Raymond A. *Linear Algebra*, Second Edition. Reading, MA: Addison-Wesley, 1990.

* Greub, Werner. *Linear Algebra*, Fourth Edition. New York, NY: Springer-Verlag, 1975.

Grossman, Stanley I. *Elementary Linear Algebra*, Fourth Edition. Philadelphia, PA: Saunders College, 1980, 1991.

Jacob, Bill. *Linear Algebra.* New York, NY: W.H. Freeman, 1990.

* Kumpel, P.G. and Thorpe, John A. *Linear Algebra with Applications to Differential Equations.* Philadelphia, PA: Saunders College, 1983.

Lang, Serge. *Introduction to Linear Algebra*, Second Edition. New York, NY: Springer-Verlag, 1986.

Leon, Steven J. *Linear Algebra with Applications*, Third Edition. New York, NY: Macmillan, 1980, 1986.

Lipschutz, Seymour. *Schaum's Solved Problems Series: 3000 Solved Problems in Linear Algebra.* New York, NY: McGraw-Hill, 1989.

* Noble, Ben and Daniel, James W. *Applied Linear Algebra*, Third Edition. Englewood Cliffs, NJ: Prentice Hall, 1969, 1988.

O'Nan, Michael and Enderton, Herbert B. *Linear Algebra*, Third Edition. San Diego, CA: Harcourt Brace Jovanovich, 1990.

* Rorres, Chris and Anton, Howard. *Applications of Linear Algebra*, Third Edition. New York, NY: John Wiley, 1977, 1984.

Rothenberg, Ronald I. *Linear Algebra with Computer Applications.* New York, NY: John Wiley, 1983.

Smith, Larry. *Linear Algebra.* New York, NY: Springer-Verlag, 1978.

*** Strang, Gilbert. *Linear Algebra and Its Applications*, Third Edition. New York, NY: Academic Press, 1976; San Diego, CA: Harcourt Brace Jovanovich, 1988.

Towers, David A. *Guide to Linear Algebra.* Houndmills, England: Macmillan Education, 1988.

* Tucker, Alan. *A Unified Introduction to Linear Algebra: Models, Methods, and Theory.* New York, NY: Macmillan, 1988.

12.2 Advanced

* Brown, William C. *A Second Course in Linear Algebra.* New York, NY: John Wiley, 1988.

Dieudonné, Jean. *Linear Algebra and Geometry.* Boston, MA: Houghton Mifflin, 1969.

* Gel'fand, Israel M. *Lectures on Linear Algebra.* Mineola, NY: Dover, 1989.

Gohberg, Israel; Lancaster, Peter; and Rodman, L. *Invariant Subspaces of Matrices with Applications.* New York, NY: John Wiley, 1986.

*** Halmos, Paul R. *Finite-Dimensional Vector Spaces.* New York, NY: Springer-Verlag, 1968, 1974.

** Herstein, I.N. and Winter, David J. *Matrix Theory and Linear Algebra.* New York, NY: Macmillan, 1968, 1988.

*** Hoffman, Kenneth and Kunze, Ray. *Linear Algebra*, Second Edition. Englewood Cliffs, NJ: Prentice Hall, 1971.

JÄRVINEN, RICHARD D. *Finite and Infinite Dimensional Linear Spaces: A Comparative Study in Algebraic and Analytic Settings.* New York, NY: Marcel Dekker, 1981.

* KAPLANSKY, IRVING. *Linear Algebra and Geometry: A Second Course.* Needham Heights, MA: Allyn and Bacon, 1969; New York, NY: Chelsea, 1974.

** LANG, SERGE. *Linear Algebra,* Third Edition. New York, NY: Springer-Verlag, 1987.

SHILOV, G.E. *Linear Algebra.* Englewood Cliffs, NJ: Prentice Hall, 1971.

12.3 Matrix Theory

** BERMAN, ABRAHAM AND PLEMMONS, ROBERT J. *Nonnegative Matrices in the Mathematical Sciences.* New York, NY: Academic Press, 1979.

* DAVIS, PHILIP J. *Circulant Matrices.* New York, NY: John Wiley, 1979.

GANTMACHER, FELIX. *Matrix Theory,* 2 Vols. New York, NY: Chelsea, 1959.

GRAYBILL, FRANKLIN A. *Introduction to Matrices with Applications in Statistics.* Belmont, CA: Wadsworth, 1969.

* HORN, ROGER A. AND JOHNSON, CHARLES R. *Matrix Analysis.* New York, NY: Cambridge University Press, 1985.

HOUSEHOLDER, ALSTON S. *The Theory of Matrices in Numerical Analysis.* New York, NY: Blaisdell, 1964.

IOHVIDOV, I.S. *Hankel and Toeplitz Matrices and Forms: Algebraic Theory.* New York, NY: Birkhäuser, 1982.

JOHNSON, CHARLES R., ED. *Matrix Theory and Applications.* Providence, RI: American Mathematical Society, 1990.

** LANCASTER, PETER AND TISMENETSKY, MIRON. *The Theory of Matrices,* Second Edition with Applications. New York, NY: Academic Press, 1985.

* MARCUS, MARVIN AND MINC, HENRYK. *Survey of Matrix Theory and Matrix Inequalities.* Boston, MA: Allyn and Bacon, 1964.

* MINC, HENRYK. *Nonnegative Matrices.* New York, NY: John Wiley, 1988.

MUIR, THOMAS. *A Treatise on the Theory of Determinants.* Mineola, NY: Dover, 1933.

NEWMAN, MORRIS. *Integral Matrices.* New York, NY: Academic Press, 1972.

PERLIS, SAM. *Theory of Matrices.* Reading, MA: Addison-Wesley, 1952.

PULLMAN, N.J. *Matrix Theory and its Applications: Selected Topics.* New York, NY: Marcel Dekker, 1976.

WEDDERBURN, J.H.M. *Lectures on Matrices.* Mineola, NY: Dover, 1934, 1964.

12.4 Special Topics

* CAMPBELL, S.L. AND MEYER, C.D., JR. *Generalized Inverses of Linear Transformations.* Brooklyn, NY: Pitman, 1979.

GOODBODY, A.M. *Cartesian Tensors: With Applications to Mechanics, Fluid Mechanics and Elasticity.* New York, NY: Halsted Press, 1982.

** GREUB, WERNER. *Multilinear Algebra,* Second Edition. New York, NY: Springer-Verlag, 1978.

ROMAN, STEVEN. *The Umbral Calculus.* New York, NY: Academic Press, 1984.

13 Algebra

13.1 Introductory Surveys

ALLENBY, R.B.J.T. *Rings, Fields and Groups: An Introduction to Abstract Algebra.* New York, NY: Edward Arnold, 1983.

** ARTIN, MICHAEL. *Algebra.* Englewood Cliffs, NJ: Prentice Hall, 1991.

BHATTACHARYA, P.B.; JAIN, S.K.; AND NAGPAUL, S.R. *Basic Abstract Algebra.* New York, NY: Cambridge University Press, 1986.

*** BIRKHOFF, GARRETT AND MAC LANE, SAUNDERS. *A Survey of Modern Algebra,* Fourth Edition. New York, NY: Macmillan, 1965, 1977.

BURNSIDE, WILLIAM SNOW AND PANTON, ARTHUR WILLIAM. *The Theory of Equations with an Introduction to the Theory of Binary Algebraic Forms,* 2 Vols. Mineola, NY: Dover, 1960.

BURTON, D. *Abstract Algebra.* Dubuque, IA: William C. Brown, 1988.

* CHILDS, LINDSAY. *A Concrete Introduction to Higher Algebra.* New York, NY: Springer-Verlag, 1979.

DEAN, RICHARD A. *Classical Abstract Algebra.* New York, NY: Harper and Row, 1990.

* FRALEIGH, JOHN B. *A First Course in Abstract Algebra,* Fourth Edition. Reading, MA: Addison-Wesley, 1976, 1989.

** GALLIAN, JOSEPH A. *Contemporary Abstract Algebra,* Second Edition. Lexington, MA: D.C. Heath, 1986, 1990.

GOLDSTEIN, LARRY J. *Abstract Algebra: A First Course.* Englewood Cliffs, NJ: Prentice Hall, 1973.

* HERSTEIN, I.N. *Abstract Algebra,* Second Edition. New York, NY: Macmillan, 1986, 1990.

HILLMAN, ABRAHAM P. AND ALEXANDERSON, GERALD L. *A First Undergraduate Course in Abstract Algebra,* Fourth Edition. Belmont, CA: Wadsworth, 1973, 1988.

* HUNGERFORD, THOMAS W. *Abstract Algebra: An Introduction.* Philadelphia, PA: Saunders College, 1990.

KOSTRIKIN, A.I. *Introduction to Algebra.* New York, NY: Springer-Verlag, 1982.

LANG, SERGE. *Undergraduate Algebra.* New York, NY: Springer-Verlag, 1987.

MARCUS, MARVIN. *Introduction to Modern Algebra.* New York, NY: Marcel Dekker, 1978.

MCCOY, NEAL H. AND JANUSZ, GERALD J. *Introduction to Modern Algebra,* Fourth Edition. Boston, MA: Allyn and Bacon, 1960, 1987.

PINTER, CHARLES C. *A Book of Abstract Algebra.* New York, NY: McGraw-Hill, 1982.

13.2 Constructive and Computational Algebra

BARBEAU, EDWARD J. *Polynomials.* New York, NY: Springer-Verlag, 1989.

CONNELL, IAN. *Modern Algebra: A Constructive Approach.* New York, NY: Elsevier Science, 1982.

* DOBBS, DAVID E. AND HANKS, ROBERT. *A Modern Course on the Theory of Equations.* Washington, NJ: Polygonal, 1980.

* HUMPHREYS, J.F. AND PREST, M.Y. *Numbers, Groups, and Codes.* New York, NY: Cambridge University Press, 1989.

MINES, RAY; RICHMAN, FRED; AND RUITENBURG, WIM. *A Course in Constructive Algebra.* New York, NY: Springer-Verlag, 1988.

SIMS, CHARLES C. *Abstract Algebra: A Computational Approach.* New York, NY: John Wiley, 1984.

13.3 Applied Algebra

BIRKHOFF, GARRETT AND BARTEE, THOMAS C. *Modern Applied Algebra.* New York, NY: McGraw-Hill, 1970.

* DORNHOFF, LARRY L. AND HOHN, FRANZ E. *Applied Modern Algebra.* New York, NY: Macmillan, 1978.

LAUFER, HENRY B. *Discrete Mathematics and Applied Modern Algebra.* Boston, MA: Prindle, Weber and Schmidt, 1984.

* LIDL, RUDOLF AND PILZ, GÜNTER. *Applied Abstract Algebra.* New York, NY: Springer-Verlag, 1984.

LIPSON, JOHN D. *Elements of Algebra and Algebraic Computing.* Reading, MA: Addison-Wesley, 1981.

** MACKIW, GEORGE. *Applications of Abstract Algebra.* New York, NY: John Wiley, 1985.

13.4 Advanced Surveys

* BOURBAKI, NICOLAS. *Elements of Mathematics: Algebra,* 2 Vols. Reading, MA: Addison-Wesley, 1973; New York, NY: Springer-Verlag, 1989, 1990.

** COHN, PAUL M. *Algebra,* 2 Vols., Second Edition. New York, NY: John Wiley, 1974, 1982.

*** HERSTEIN, I.N. *Topics in Algebra,* Second Edition. New York, NY: John Wiley, 1975.

* HUNGERFORD, THOMAS W. *Algebra.* New York, NY: Springer-Verlag, 1974.

*** JACOBSON, NATHAN. *Basic Algebra I and II,* Second Edition. New York, NY: W.H. Freeman, 1974, 1989.

* JACOBSON, NATHAN. *Lectures in Abstract Algebra,* 2 Vols. New York, NY: Springer-Verlag, 1953, 1975.

KOSTRIKIN, A.I. AND SHAFAREVICH, IGOR R., EDS. *Algebra I: Basic Notions of Algebra.* New York, NY: Springer-Verlag, 1990.

* LANG, SERGE. *Algebra,* Second Edition. Reading, MA: Addison-Wesley, 1965, 1984.

LEDERMANN, WALTER AND VAJDA, STEVEN, EDS. *Algebra.* Handbook of Applicable Mathematics, Volume I. New York, NY: John Wiley, 1980.

*** MAC LANE, SAUNDERS AND BIRKHOFF, GARRETT. *Algebra,* Third Edition. New York, NY: Macmillan, 1967; New York, NY: Chelsea, 1988.

*** VAN DER WAERDEN, B.L. *Algebra,* 2 Vols., Seventh Edition. (Original title: *Modern Algebra.*) New York, NY: Frederick Ungar, 1950; New York, NY: Springer-Verlag, 1991.

13.5 Group Theory

ASCHBACHER, MICHAEL. *The Finite Simple Groups and Their Classification.* New Haven, CT: Yale University Press, 1980.

** BUDDEN, F.J. *The Fascination of Groups.* New York, NY: Cambridge University Press, 1972.

* BURN, R.P. *Groups: A Path to Geometry.* New York, NY: Cambridge University Press, 1985, 1987.

CONWAY, JOHN HORTON, *et al. Atlas of Finite Groups: Maximal Subgroups and Ordinary Characters for Simple Groups.* New York, NY: Clarendon Press, 1985.

* CURTIS, CHARLES W. AND REINER, IRVING. *Representation Theory of Finite Groups and Associative Algebras.* New York, NY: John Wiley, 1962.

* CURTIS, CHARLES W. AND REINER, IRVING. *Methods of Representation Theory with Applications to Finite Groups and Orders.* New York, NY: John Wiley, 1981.

DIXON, J.D. *Problems in Group Theory.* New York, NY: Blaisdell, 1967.

FEIGELSTOCK, S. *Additive Groups of Rings.* Brooklyn, NY: Pitman, 1983.

FUCHS, L. *Abelian Groups.* New York, NY: Academic Press, 1970.

** GORENSTEIN, DANIEL. *Finite Simple Groups: An Introduction to Their Classification.* New York, NY: Plenum Press, 1982.

* GORENSTEIN, DANIEL. *The Classification of Finite Simple Groups.* New York, NY: Plenum Press, 1983.

GROVE, L.C. AND BENSON, C.T. *Finite Reflection Groups,* Second Edition. New York, NY: Springer-Verlag, 1985.

HALL, MARSHALL, JR. AND SENIOR, J.K. *Groups of Order* $2^n (n \leq 6)$. New York, NY: Macmillan, 1964.

*** HALL, MARSHALL, JR. *The Theory of Groups,* Second Edition. New York, NY: Macmillan, 1959; New York, NY: Chelsea, 1973.

HILL, VICTOR E. *Groups, Representations, and Characters.* New York, NY: Hafner Press, 1975.

JOHNSON, D.L. *Presentations of Groups.* New York, NY: Cambridge University Press, 1976.

* KAPLANSKY, IRVING. *Infinite Abelian Groups,* Revised Edition. Ann Arbor, MI: University of Michigan Press, 1969.

* KUROSH, ALEXANDER G. *The Theory of Groups,* 2 Vols., Second Edition. New York, NY: Chelsea, 1960, 1970.

** LEDERMANN, WALTER. *Introduction to Group Characters,* Second Edition. New York, NY: Cambridge University Press, 1977, 1987.

*** ROTMAN, JOSEPH J. *An Introduction to the Theory of Groups,* Third Edition. Needham Heights, MA: Allyn and Bacon, 1965, 1984.

SCOTT, WILLIAM R. *Group Theory.* Englewood Cliffs, NJ: Prentice Hall, 1964; Mineola, NY: Dover, 1987.

SERRE, JEAN-PIERRE. *Linear Representations of Finite Groups.* New York, NY: Springer-Verlag, 1977.

WEINSTEIN, MICHAEL. *Examples of Groups.* Washington, NJ: Polygonal, 1977.

* WEYL, HERMANN. *The Classical Groups: Their Invariants and Representatives.* Princeton, NJ: Princeton University Press, 1946.

13.6 Rings and Ideals

COHN, PAUL M. *Free Rings and Their Relations,* Second Edition. New York, NY: Academic Press, 1985.

* GOODEARL, K.R. AND WARFIELD, R.B., JR. *An Introduction to Noncommutative Noetherian Rings.* New York, NY: Cambridge University Press, 1989.

*** HERSTEIN, I.N. *Non-Commutative Rings.* Washington, DC: Mathematical Association of America, 1968.

* HERSTEIN, I.N. *Rings with Involution.* Chicago, IL: University of Chicago Press, 1976.

* JACOBSON, NATHAN. *The Structure of Rings,* Revised Edition. Providence, RI: American Mathematical Society, 1964.

JANS, J. *Rings and Homology.* New York, NY: Holt, Rinehart and Winston, 1964.

** KAPLANSKY, IRVING. *Fields and Rings,* Revised Second Edition. Chicago, IL: University of Chicago Press, 1969, 1974.

KOSTRIKIN, A.I. AND SHAFAREVICH, IGOR R., EDS. *Algebra II: Non-Commutative Rings, Identities.* New York, NY: Springer-Verlag, 1991.

* LAMBEK, JOACHIM. *Lectures on Rings and Modules,* Second Edition. New York, NY: Blaisdell, 1966; New York, NY: Chelsea, 1976.

* MCCONNELL, J.C. AND ROBSON, J.C. *Noncommutative Noetherian Rings.* New York, NY: John Wiley, 1988.

MCCOY, NEAL H. *Rings and Ideals.* Washington, DC: Mathematical Association of America, 1948.

MCCOY, NEAL H. *The Theory of Rings.* New York, NY: Macmillan, 1964; New York, NY: Chelsea, 1973.

PASSMAN, DONALD S. *The Algebraic Structure of Group Rings.* Melbourne, FL: Robert E. Krieger, 1985.

* ROBINSON, ABRAHAM. *Numbers and Ideals.* San Francisco, CA: Holden-Day, 1965.

* ROWEN, LOUIS HALLE. *Ring Theory,* 2 Vols. New York, NY: Academic Press, 1988.

ROWEN, LOUIS HALLE. *Polynomial Identities in Ring Theory.* New York, NY: Academic Press, 1980.

SHARPE, DAVID. *Rings and Factorization.* New York, NY: Cambridge University Press, 1987.

SMALL, LANCE W. *Noetherian Rings and Their Applications.* Providence, RI: American Mathematical Society, 1987.

STENSTRÖM, BO. *Rings of Quotients: An Introduction to Methods of Ring Theory.* New York, NY: Springer-Verlag, 1975.

13.7 Fields and Galois Theory

* ADAMSON, IAIN T. *Introduction to Field Theory,* Second Edition. New York, NY: Cambridge University Press, 1982.

* ARTIN, EMIL. *Galois Theory,* Second Revised Edition. Notre Dame, IN: University of Notre Dame Press, 1966.

BRAWLEY, JOEL V. AND SCHNIBBEN, GEORGE E. *Infinite Algebraic Extensions of Finite Fields.* Providence, RI: American Mathematical Society, 1989.

* EDWARDS, HAROLD M. *Galois Theory.* New York, NY: Springer-Verlag, 1984.

** GAAL, LISL. *Classical Galois Theory with Examples,* Fourth Edition. Boston, MA: Markham, 1971; New York, NY: Chelsea, 1973, 1988.

GARLING, D.J.H. *A Course in Galois Theory.* New York, NY: Cambridge University Press, 1986.

*** HADLOCK, CHARLES R. *Field Theory and Its Classical Problems.* Washington, DC: Mathematical Association of America, 1978.

LANG, SERGE. *Cyclotomic Fields I and II,* Second Edition. New York, NY: Springer-Verlag, 1978–80, 1990.

* LIDL, RUDOLF AND NIEDERREITER, HARALD. *Introduction to Finite Fields and Their Applications.* New York, NY: Cambridge University Press, 1986.

* LIEBER, LILLIAN R. *Galois and the Theory of Groups: A Bright Star in Mathesis.* Brooklyn, NY: Galois Institute of Mathematics and Art, 1961.

* MCCARTHY, PAUL J. *Algebraic Extensions of Fields,* Second Edition. New York, NY: Chelsea, 1976.

ROTMAN, JOSEPH J. *Galois Theory.* New York, NY: Springer-Verlag, 1990.

*** STEWART, IAN. *Galois Theory,* Second Edition. New York, NY: Halsted Press, 1973; New York, NY: Chapman and Hall, 1989.

13.8 Commutative Algebra

** ATIYAH, MICHAEL F. AND MACDONALD, I.G. *Introduction to Commutative Algebra.* Reading, MA: Addison-Wesley, 1969.

* BOURBAKI, NICOLAS. *Elements of Mathematics: Commutative Algebra,* Reading, MA: Addison-Wesley, 1972; New York, NY: Springer-Verlag, 1989.

HUTCHINS, HARRY C. *Examples of Commutative Rings.* Washington, NJ: Polygonal, 1981.

* KAPLANSKY, IRVING. *Commutative Rings,* Revised Edition. Chicago, IL: University of Chicago Press, 1974; Boston, MA: Allyn and Bacon, 1974.

KUNZ, E. *Introduction to Commutative Algebra and Algebraic Geometry.* New York, NY: Birkhäuser, 1985.

MATSUMURA, HIDEYUKI. *Commutative Ring Theory.* New York, NY: Cambridge University Press, 1986.

* NAGATA, MASAYOSHI. *Local Rings.* New York, NY: Interscience, 1962.

*** ZARISKI, OSCAR AND SAMUEL, PIERRE. *Commutative Algebra,* 2 Vols. New York, NY: Springer-Verlag, 1975, 1976.

13.9 Homological Algebra

GERAMITA, ANTHONY V. AND SMALL, CHARLES. *Introduction to Homological Methods in Commutative Rings.* Kingston: Queen's University Press, 1976.

* HILTON, PETER J. AND STAMMBACH, U. *A Course in Homological Algebra.* New York, NY: Springer-Verlag, 1971.

** MAC LANE, SAUNDERS. *Homology.* New York, NY: Springer-Verlag, 1963.

NORTHCOTT, D.G. *A First Course of Homological Algebra.* New York, NY: Cambridge University Press, 1973, 1980.

** ROTMAN, JOSEPH J. *An Introduction to Homological Algebra.* New York, NY: Academic Press, 1979.

13.10 Category Theory

* * BARR, MICHAEL; AND WELLS, CHARLES. *Category Theory for Computing Science.* Hemel Hemstead, UK: Prentice Hall International, 1990.
* BLYTH, T.S. *Categories.* White Plains, NY: Longman, 1986.
* * HERRLICH, HORST AND STRECKER, GEORGE E. *Category Theory: An Introduction,* Second Edition. Boston, MA: Allyn and Bacon, 1973; Berlin: Heldermann Verlag, 1979.
* LAMBEK, JOACHIM AND SCOTT, P.J. *Introduction to Higher Order Categorical Logic.* New York, NY: Cambridge University Press, 1986, 1988.
* ** MAC LANE, SAUNDERS. *Categories for the Working Mathematician.* New York, NY: Springer-Verlag, 1971.
* PAREIGIS, B. *Categories and Functors.* New York, NY: Academic Press, 1970.

13.11 Lie Algebras

* HUMPHREYS, JAMES E. *Introduction to Lie Algebras and Representation Theory.* New York, NY: Springer-Verlag, 1972.
* * JACOBSON, NATHAN. *Lie Algebras.* New York, NY: John Wiley, 1962; Mineola, NY: Dover, 1979.
* * KAPLANSKY, IRVING. *Lie Algebras and Locally Compact Groups.* Chicago, IL: University of Chicago Press, 1971.
* * SAMELSON, HANS. *Notes on Lie Algebras.* New York, NY: Van Nostrand Reinhold, 1969; New York, NY: Springer-Verlag, 1990.
* WINTER, DAVID J. *Abstract Lie Algebras.* Cambridge, MA: MIT Press, 1972.

13.12 Universal Algebra

* BURRIS, STANLEY AND SANKAPPANAVAR, H.P. *A Course in Universal Algebra.* New York, NY: Springer-Verlag, 1981.
* * COHN, PAUL M. *Universal Algebra.* Norwell, MA: D. Reidel, 1981.
* * GRÄTZER, GEORGE. *Universal Algebra,* Second Edition. New York, NY: Springer-Verlag, 1979.

13.13 Special Topics

* ARTIN, EMIL. *Geometric Algebra.* New York, NY: John Wiley, 1957.
* * BACHMANN, FRIEDRICH; SCHMIDT, ECKART; AND GARNER, CYRIL W.L. *n-gons.* Toronto: University of Toronto Press, 1975.
* BASS, HYMAN. *Algebraic K-Theory.* Redwood City, CA: Benjamin Cummings, 1968.
* * HALMOS, PAUL R. *Lectures on Boolean Algebras.* New York, NY: Springer-Verlag, 1974.
* MELDRUM, J.D.P. *Near-rings and Their Links with Groups.* Brooklyn, NY: Pitman, 1985.
* MONK, J. DONALD AND BONNET, ROBERT, EDS. *Handbook of Boolean Algebras,* 3 Vols. Amsterdam: North-Holland, 1989.
* * MONTGOMERY, SUSAN, *et al.*, EDS. *Selected Papers on Algebra.* Washington, DC: Mathematical Association of America, 1977.
* SIKORSKI, R. *Boolean Algebras.* New York, NY: Springer-Verlag, 1960.
* SILVESTER, JOHN R. *Introduction to Algebraic K-Theory.* New York, NY: Chapman and Hall, 1981.

14 Geometry

14.1 General

*** Banchoff, Thomas F. *Beyond the Third Dimension: Geometry, Computer Graphics, and Higher Dimensions.* New York, NY: Scientific American Library, 1990.

Blackwell, William. *Geometry in Architecture.* New York, NY: John Wiley, 1984.

Bold, Benjamin. *Famous Problems of Geometry and How to Solve Them.* New York, NY: Van Nostrand Reinhold, 1969; Mineola, NY: Dover, 1982.

Burger, Dionys. *Sphereland.* New York, NY: Thomas Y. Crowell, 1965.

** Croft, Hallard T.; Falconer, Kenneth J.; and Guy, Richard K. *Unsolved Problems in Geometry.* New York, NY: Springer-Verlag, 1991.

* Fischer, Gerd, ed. *Mathematical Models from the Collections of Universities and Museums,* 2 Vols. Wiesbaden: Friedr. Vieweg and Sohn, 1986.

** Friedrichs, Kurt Otto. *From Pythagoras to Einstein.* Washington, DC: Mathematical Association of America, 1965.

*** Hilbert, David and Cohn-Vossen, S. *Geometry and the Imagination.* New York, NY: Chelsea, 1952.

*** Hildebrandt, Stefan and Tromba, Anthony J. *Mathematics and Optimal Form.* New York, NY: Scientific American Library, 1984.

Ivins, William M., Jr. *Art and Geometry: A Study in Space Intuitions.* Mineola, NY: Dover, 1964.

** Krause, Eugene F. *Taxicab Geometry: An Adventure in Non-Euclidean Geometry.* Reading, MA: Addison-Wesley, 1975; Mineola, NY: Dover, 1986.

Lord, E.A. and Wilson, C.B. *The Mathematical Description of Shape and Form.* New York, NY: Halsted Press, 1984.

March, Lionel and Steadman, Philip. *The Geometry of Environment: An Introduction to Spatial Organization in Design.* Cambridge, MA: MIT Press, 1974.

Pottage, John. *Geometrical Investigations: Illustrating the Art of Discovery in the Mathematical Field.* Reading, MA: Addison-Wesley, 1983.

* Row, T. Sundara. *Geometric Exercises in Paper Folding.* Mineola, NY: Dover, 1966.

Rucker, Rudy. *Geometry, Relativity, and the Fourth Dimension.* Mineola, NY: Dover, 1977.

* Stewart, Bonnie M. *Adventures Among the Toroids,* Second Edition. 4494 Wasau Road, Okemos, MI: 1970, 1980.

14.2 Surveys

* Berger, Marcel. *Geometry,* 2 Vols. New York, NY: Springer-Verlag, 1987.

* Berger, Marcel, *et al. Problems in Geometry.* New York, NY: Springer-Verlag, 1984.

* Cederberg, Judith N. *A Course in Modern Geometries.* New York, NY: Springer-Verlag, 1989.

Court, Nathan A. *College Geometry,* Second Edition. New York, NY: Barnes and Noble, 1952.

Cox, Philip L. *Geometry in Easy Steps: An Informal Approach.* Needham Heights, MA: Allyn and Bacon, 1983.

*** Coxeter, H.S.M. *Introduction to Geometry,* Second Edition. New York, NY: John Wiley, 1969.

*** Eves, Howard W. *A Survey of Geometry,* Second Revised Edition. Boston, MA: Allyn and Bacon, 1972.

Lial, Margaret L.; Steffensen, Arnold R.; and Johnson, L. Murphy. *Essentials of Geometry for College Students.* Glenview, IL: Scott Foresman, 1990.

* Melzak, Z.A. *Invitation to Geometry.* New York, NY: John Wiley, 1983.

* Millman, Richard S. and Parker, George D. *Geometry: A Metric Approach with Models,* Second Edition. New York, NY: Springer-Verlag, 1981, 1991.

Pedoe, Dan. *Geometry: A Comprehensive Course.* (Former title: *A Course of Geometry for Colleges and Universities.*) New York, NY: Cambridge University Press, 1970; Mineola, NY: Dover, 1988.

* PEDOE, DAN. *Geometry and the Visual Arts.* (Former title: *Geometry and the Liberal Arts.*) New York, NY: St. Martin's Press, 1978; Mineola, NY: Dover, 1983.

POSTNIKOV, MIKHAÏL. *Lectures in Geometry,* 2 Vols. Moscow: MIR, 1982, 1986.

PRENOWITZ, WALTER AND JORDAN, M. *Basic Concepts of Geometry.* New York, NY: John Wiley, 1965.

SMART, JAMES R. *Modern Geometries,* Third Edition. Pacific Grove, CA: Brooks/Cole, 1973, 1988.

** STEHNEY, ANN K., *et al.*, EDS. *Selected Papers on Geometry.* Washington, DC: Mathematical Association of America, 1979.

TULLER, ANITA. *A Modern Introduction to Geometries.* New York, NY: Van Nostrand Reinhold, 1967.

14.3 School Geometry

* BRUNI, JAMES V. *Experiencing Geometry.* Belmont, CA: Wadsworth, 1977.

CLEMENS, STANLEY R.; O'DAFFER, PHARES G.; AND CLOONEY, THOMAS J. *Geometry.* Reading, MA: Addison-Wesley, 1983.

* FETISOV, A.I. *Proof in Geometry.* Moscow: MIR, 1978.

HOFFER, ALAN. *Geometry.* Reading, MA: Addison-Wesley, 1979.

** JACOBS, HAROLD R. *Geometry,* Second Edition. New York, NY: W.H. Freeman, 1974, 1986.

KEMPE, A.B. *How to Draw a Straight Line.* Reston, VA: National Council of Teachers of Mathematics, 1977.

* KONKLE, GAIL S. *Shapes and Perceptions: An Intuitive Approach to Geometry.* Boston, MA: Prindle, Weber and Schmidt, 1974.

LOOMIS, E. *The Pythagorean Proposition.* Reston, VA: National Council of Teachers of Mathematics, 1968.

MOISE, EDWIN E. AND DOWNS, FLOYD L. *Geometry.* Reading, MA: Addison-Wesley, 1975.

** O'DAFFER, PHARES G. AND CLEMENS, STANLEY R. *Geometry: An Investigative Approach.* Reading, MA: Addison-Wesley, 1976.

14.4 Euclidean and Non-Euclidean Geometry

BESKIN, N.M. *Dividing a Segment in a Given Ratio.* Moscow: MIR, 1975.

*** COXETER, H.S.M., *et al. Geometry Revisited.* Washington, DC: Mathematical Association of America, 1967.

* COXETER, H.S.M. *Non-Euclidean Geometry.* Toronto: University of Toronto Press, 1957.

* DUDLEY, UNDERWOOD. *A Budget of Trisections.* New York, NY: Springer-Verlag, 1987.

* FABER, RICHARD L. *Foundations of Euclidean and Non-Euclidean Geometry.* New York, NY: Marcel Dekker, 1983.

* GANS, DAVID. *An Introduction to Non-Euclidean Geometry.* New York, NY: Academic Press, 1973.

** GRAY, JEREMY. *Ideas of Space: Euclidean, Non-Euclidean, and Relativistic,* Second Edition. New York, NY: Clarendon Press, 1979, 1989.

* GREENBERG, MARVIN JAY. *Euclidean and Non-Euclidean Geometries: Development and History.* New York, NY: W.H. Freeman, 1974, 1980.

*** HILBERT, DAVID. *The Foundations of Geometry,* Tenth Edition. La Salle, IL: Open Court, 1962, 1971.

KELLY, PAUL J. AND MATTHEWS, GORDON. *The Non-Euclidean, Hyperbolic Plane: Its Structure and Consistency.* New York, NY: Springer-Verlag, 1981.

KOGAN, B. YU. *The Application of Mechanics to Geometry.* Chicago, IL: University of Chicago Press, 1974.

LOBACHEVSKI, N. *The Theory of Parallels.* Peru, IL: Open Court, 1914.

* MARTIN, GEORGE E. *The Foundations of Geometry and the Non-Euclidean Plane.* New York, NY: Springer-Verlag, 1975.

MOISE, EDWIN E. *Elementary Geometry from an Advanced Standpoint*, Second Edition. Reading, MA: Addison-Wesley, 1974.

MORLEY, F. AND MORLEY, F.M. *Inversive Geometry*. New York, NY: Chelsea, 1954.

NIKULIN, V.V. AND SHAFAREVICH, IGOR R. *Geometries and Groups*. New York, NY: Springer-Verlag, 1987.

PEDOE, DAN. *Circles: A Mathematical View*. Mineola, NY: Dover, 1979.

POSAMENTIER, ALFRED S. *Excursions in Advanced Euclidean Geometry*. Reading, MA: Addison-Wesley, 1984.

ROSENFELD, B.A. AND SERGEEVA, N.D. *Stereographic Projection*. Moscow: MIR, 1977.

RYAN, PATRICK J. *Euclidean and Non-Euclidean Geometry: An Analytical Approach*. New York, NY: Cambridge University Press, 1986.

SCHWERDTFEGER, HANS. *Geometry of Complex Numbers: Circle Geometry, Moebius Transformation, Non-Euclidean Geometry*. Mineola, NY: Dover, 1979.

SHREIDER, YU. A. *What is Distance?* Chicago, IL: University of Chicago Press, 1974.

* SZMIELEW, WANDA. *From Affine to Euclidean Geometry: An Axiomatic Approach*. Norwell, MA: D. Reidel, 1983.

TRUDEAU, RICHARD J. *The Non-Euclidean Revolution*. New York, NY: Birkhäuser, 1987.

VAISMAN, IZU. *Foundations of Three-Dimensional Euclidean Geometry*. New York, NY: Marcel Dekker, 1980.

YAGLOM, I.M. *A Simple Non-Euclidean Geometry and Its Physical Basis*. New York, NY: Springer-Verlag, 1979.

* YATES, ROBERT C. *The Trisection Problem*. Reston, VA: National Council of Teachers of Mathematics, 1971.

14.5 Polyhedra, Tilings, Symmetry

** ARMSTRONG, M.A. *Groups and Symmetry*. New York, NY: Springer-Verlag, 1988.

* BAGLIVO, JENNY A. AND GRAVER, JACK E. *Incidence and Symmetry in Design and Architecture*. New York, NY: Cambridge University Press, 1983.

BESKIN, N.M. *Images of Geometric Solids*. Moscow: MIR, 1985.

* BUNCH, BRYAN H. *Reality's Mirrors: Exploring the Mathematics of Symmetry*. New York, NY: John Wiley, 1989.

* COXETER, H.S.M., *et al.*, EDS. *M.C. Escher: Art and Science*. New York, NY: Elsevier Science, 1986.

COXETER, H.S.M. *Regular Polytopes*, Third Edition. New York, NY: Macmillan, 1963; Mineola, NY: Dover, 1973.

** CUNDY, M.H. AND ROLLETT, A.P. *Mathematical Models*. New York, NY: Clarendon Press, 1961.

*** GRÜNBAUM, BRANKO AND SHEPHARD, G.C. *Tilings and Patterns*. New York, NY: W.H. Freeman, 1986, 1989.

* GRÜNBAUM, BRANKO. *Convex Polytopes*. New York, NY: John Wiley, 1967.

HARGITTAI, ISTVAN, ED. *Symmetry: Unifying Human Understanding*. Elmsford, NY: Pergamon Press, 1986.

* HENDERSON, LINDA D. *The Fourth Dimension and Non-Euclidean Geometry in Modern Art*. Princeton, NJ: Princeton University Press, 1983.

* HILTON, PETER J. AND PEDERSEN, JEAN. *Build Your Own Polyhedra*. Reading, MA: Addison-Wesley, 1988.

HOLDEN, ALAN. *Orderly Tangles: Cloverleafs, Gordian Knots, and Regular Polylinks*. New York, NY: Columbia University Press, 1983.

* HOLDEN, ALAN. *Shapes, Space, and Symmetry*. New York, NY: Columbia University Press, 1971.

KAVANAU, J. LEE. *Symmetry: An Analytical Treatment*. West Los Angeles, CA: Science Software Systems, 1980.

* LOCKWOOD, E.H. AND MACMILLAN, R.H. *Geometric Symmetry.* New York, NY: Cambridge University Press, 1978.

LOEB, ARTHUR L. *Color and Symmetry.* New York, NY: John Wiley, 1971.

LOEB, ARTHUR L. *Space Structures: Their Harmony and Counterpoint.* Reading, MA: Addison-Wesley, 1976.

LYNDON, ROGER C. *Groups and Geometry.* New York, NY: Cambridge University Press, 1985.

MACGILLAVRY, CAROLINE H. *Fantasy & Symmetry: The Periodic Drawings of M.S. Escher.* New York, NY: Harry N. Abrams, 1976.

* MARTIN, GEORGE E. *Transformation Geometry: An Introduction to Symmetry.* New York, NY: Springer-Verlag, 1982.

PEARCE, PETER AND PEARCE, SUSAN. *Polyhedra Primer.* New York, NY: Van Nostrand Reinhold, 1978.

PUGH, ANTHONY. *Polyhedra: A Visual Approach.* Berkeley, CA: University of California Press, 1976.

RANUCCI, E.R. AND TEETERS, J.L. *Creating Escher-Type Drawings.* Palo Alto, CA: Creative Publ., 1977.

* ROSEN, JOE. *Symmetry Discovered: Concepts and Applications in Nature and Science.* New York, NY: Cambridge University Press, 1975.

** SCHATTSCHNEIDER, DORIS J. *Visions of Symmetry: Notebooks, Periodic Drawings, and Related Work of M.C. Escher.* New York, NY: W.H. Freeman, 1990.

* SENECHAL, MARJORIE AND FLECK, GEORGE, EDS. *Patterns of Symmetry.* Amherst, MA: University of Massachusetts Press, 1977.

** SENECHAL, MARJORIE AND FLECK, GEORGE, EDS. *Shaping Space: A Polyhedral Approach.* New York, NY: Birkhäuser, 1988.

SHUBNIKOV, A.V. AND KOPTSIK, V.A. *Symmetry in Science and Art.* New York, NY: Plenum Press, 1974.

STEVENS, PETER S. *Patterns in Nature.* Waltham, MA: Little, Brown, 1974.

* TÓTH, L. FEJES. *Regular Figures.* Elmsford, NY: Pergamon Press, 1964.

* WENNINGER, MAGNUS J. *Polyhedron Models.* New York, NY: Cambridge University Press, 1971.

WENNINGER, MAGNUS J. *Spherical Models.* New York, NY: Cambridge University Press, 1979.

*** WEYL, HERMANN. *Symmetry.* Princeton, NJ: Princeton University Press, 1952.

14.6 Geometric Transformations

* BAKEL'MAN, I. YA. *Inversions.* Chicago, IL: University of Chicago Press, 1974.

ECCLES, FRANK M. *An Introduction to Transformational Geometry.* Reading, MA: Addison-Wesley, 1971.

* GANS, DAVID. *Transformations and Geometries.* Englewood Cliffs, NJ: Appleton-Century-Crofts, 1969.

** YAGLOM, I.M. *Geometric Transformations,* 3 Vols. Washington, DC: Mathematical Association of America, 1962–73.

14.7 Projective Geometry

* BLUMENTHAL, LEONARD M. *Modern View of Geometry.* New York, NY: W.H. Freeman, 1961; Mineola, NY: Dover, 1980.

** COXETER, H.S.M. *Projective Geometry,* Second Edition. Toronto: University of Toronto Press, 1974.

* COXETER, H.S.M. *The Real Projective Plane.* New York, NY: Cambridge University Press, 1961.

DORWART, HAROLD L. *Geometry of Incidence.* Englewood Cliffs, NJ: Prentice Hall, 1966.

FISHBACK, WILLIAM T. *Projective and Euclidean Geometry,* Second Edition. New York, NY: John Wiley, 1969.

GARNER, LYNN E. *An Outline of Projective Geometry.* Amsterdam: North-Holland, 1981.

HIRSCHFELD, J.W.P. *Finite Projective Spaces of Three Dimensions.* New York, NY: Clarendon Press, 1985.

HIRSCHFELD, J.W.P. *Projective Geometries Over Finite Fields.* New York, NY: Clarendon Press, 1979.

PENNA, MICHAEL A. AND PATTERSON, RICHARD R. *Projective Geometry and Its Applications to Computer Graphics.* Englewood Cliffs, NJ: Prentice Hall, 1986.

SAMUEL, PIERRE. *Projective Geometry.* New York, NY: Springer-Verlag, 1988.

* VEBLEN, OSWALD AND YOUNG, JOHN WESLEY. *Projective Geometry,* 2 Vols. New York, NY: Blaisdell, 1938, 1946.

WYLIE, CLARENCE R. *Introduction to Projective Geometry.* New York, NY: McGraw-Hill, 1970.

* YOUNG, JOHN WESLEY. *Projective Geometry.* Washington, DC: Mathematical Association of America, 1930.

14.8 Computational Geometry

** ABELSON, HAROLD AND DISESSA, ANDREA A. *Turtle Geometry: The Computer as a Medium for Exploring Mathematics.* Cambridge, MA: MIT Press, 1981.

BILLSTEIN, RICK; LIBESKIND, S.; AND LOTT, JOHNNY W. *Logo.* Redwood City, CA: Benjamin Cummings, 1985.

CLAYSON, J. *Visual Modeling with Logo.* Cambridge, MA: MIT Press, 1988.

EDELSBRUNNER, H. *Algorithms in Computational Geometry.* New York, NY: Springer-Verlag, 1987.

** O'ROURKE, JOSEPH. *Art Gallery Theorems and Algorithms.* New York, NY: Oxford University Press, 1987.

* PREPARATA, FRANCO P. AND SHAMOS, MICHAEL I. *Computational Geometry: An Introduction.* New York, NY: Springer-Verlag, 1985.

TOUSSAINT, G., ED. *Computational Geometry.* Amsterdam: North-Holland, 1985.

14.9 Discrete and Combinatorial Geometry

AMBARTZUMIAN, R.V. *Combinatorial Integral Geometry with Applications to Mathematical Stereology.* New York, NY: John Wiley, 1982.

BATTEN, LYNN MARGARET. *Combinatorics of Finite Geometries.* New York, NY: Cambridge University Press, 1986.

* BOLTYANSKII, VLADIMIR G. AND GOHBERG, ISRAEL. *Results and Problems in Combinatorial Geometry.* New York, NY: Cambridge University Press, 1985.

* BOLTYANSKII, VLADIMIR G. AND GOHBERG, ISRAEL. *The Decomposition of Figures into Smaller Parts.* Chicago, IL: University of Chicago Press, 1980.

* BOLTYANSKII, VLADIMIR G. *Equivalent and Equidecomposable Figures.* Lexington, MA: D.C. Heath, 1963.

** BOLTYANSKII, VLADIMIR G. *Hilbert's Third Problem.* Silver Spring, MD: V.H. Winston, 1978.

** BONNESEN, T. AND FENCHEL, W. *Theory of Convex Bodies.* Moscow, ID: BCS Associates, 1987.

* HADWIGER, HUGO AND DEBRUNNER, HANS. *Combinatorial Geometry in the Plane.* New York, NY: Holt, Rinehart and Winston, 1964.

** KAZARINOFF, NICHOLAS D. *Geometric Inequalities.* Washington, DC: Mathematical Association of America, 1961, 1975.

KELLY, PAUL J. AND WEISS, MAX L. *Geometry and Convexity: A Study in Mathematical Methods.* New York, NY: John Wiley, 1979.

** LAY, STEVEN R. *Convex Sets and Their Applications.* New York, NY: John Wiley, 1982.

* LEDERMANN, WALTER AND VAJDA, STEVEN, EDS. *Geometry and Combinatorics.* Handbook of Applicable Mathematics, Volume V. New York, NY: John Wiley, 1985.

** MITRINOVIĆ, DRAGOSLAV S.; PEČARIĆ, J.E.; AND VOLENEC, V. *Recent Advances in Geometric Inequalities.* Norwell, MA: Kluwer Academic, 1989.

MOSER, WILLIAM. *Problems in Discrete Geometry,* Fifth Edition. Toronto: McGill-Queen's University Press, 1979, 1980.

PRENOWITZ, WALTER AND JANTOSCIAK, JAMES. *Join Geometries: A Theory of Convex Sets and Linear Geometry.* New York, NY: Springer-Verlag, 1979.

* YAGLOM, I.M. AND BOLTYANSKII, VLADIMIR G. *Convex Figures.* New York, NY: Holt, Rinehart and Winston, 1961.

14.10 Differential Geometry

ABRAHAM, RALPH H.; MARSDEN, JERROLD E.; AND RATIU, T. *Manifolds, Tensor Analysis, and Applications.* New York, NY: Springer-Verlag, 1988.

* ARNOLD, V.I. *Catastrophe Theory,* Second Edition. New York, NY: Springer-Verlag, 1984, 1986.

BERGER, MARCEL AND GOSTIAUX, BERNARD. *Differential Geometry: Manifolds, Curves, and Surfaces.* New York, NY: Springer-Verlag, 1988.

* BOOTHBY, WILLIAM M. *An Introduction to Differentiable Manifolds and Riemannian Geometry,* Second Edition. New York, NY: Academic Press, 1975, 1986.

* BURKE, WILLIAM L. *Applied Differential Geometry.* New York, NY: Cambridge University Press, 1985.

** DO CARMO, MANFREDO P. *Differential Geometry of Curves and Surfaces.* Englewood Cliffs, NJ: Prentice Hall, 1976.

* DODSON, C.T.J. AND POSTON, TIM. *Tensor Geometry: The Geometric Viewpoint and its Uses.* Brooklyn, NY: Pitman, 1977.

FABER, RICHARD L. *Differential Geometry and Relativity Theory: An Introduction.* New York, NY: Marcel Dekker, 1983.

FLANDERS, HARLEY. *Differential Forms with Applications to the Physical Sciences.* New York, NY: Academic Press, 1963; Mineola, NY: Dover, 1989.

HOPF, HEINZ. *Differential Geometry in the Large.* New York, NY: Springer-Verlag, 1983.

** HSIUNG, CHUAN-CHIH. *A First Course in Differential Geometry.* New York, NY: John Wiley, 1981.

KAHN, DONALD W. *Introduction to Global Analysis.* New York, NY: Academic Press, 1980.

* KOBAYASHI, SHOSHICHI AND NOMIZU, KATSUMI. *Foundations of Differential Geometry,* 2 Vols. New York, NY: John Wiley, 1963–69.

LU, YUNG-CHEN. *Singularity Theory and an Introduction to Catastrophe Theory.* New York, NY: Springer-Verlag, 1976.

MARMO, GIUSEPPE, *et al. Dynamical Systems: A Differential Geometric Approach to Symmetry and Reduction.* New York, NY: John Wiley, 1985.

* MILLMAN, RICHARD S. AND PARKER, GEORGE D. *Elements of Differential Geometry.* Englewood Cliffs, NJ: Prentice Hall, 1977.

O'NEILL, BARRETT. *Elementary Differential Geometry.* New York, NY: Academic Press, 1966.

** SPIVAK, MICHAEL D. *A Comprehensive Introduction to Differential Geometry,* Second Edition, 5 Vols. Boston, MA: Publish or Perish, 1970–79.

STOKER, J.J. *Differential Geometry.* New York, NY: John Wiley, 1969.

** STRUIK, DIRK JAN. *Lectures on Classical Differential Geometry,* Second Edition. (Former title: *Differential Geometry.*) Reading, MA: Addison-Wesley, 1961; Mineola, NY: Dover, 1988.

THORPE, JOHN A. *Elementary Topics in Differential Geometry.* New York, NY: Springer-Verlag, 1979.

VEBLEN, OSWALD AND WHITEHEAD, J.H.C. *The Foundations of Differential Geometry.* New York, NY: Cambridge University Press, 1967.

14.11 Algebraic Geometry

* BRIESKORN, EGBERT AND KNÖRRER, HORST. *Plane Algebraic Curves.* New York, NY: Birkhäuser, 1986.

BRUCE, J.W. AND GIBLIN, P.J. *Curves and Singularities: A Geometrical Introduction to Singularity Theory.* New York, NY: Cambridge University Press, 1984.

CLEMENS, C. HERBERT. *A Scrapbook of Complex Curve Theory.* New York, NY: Plenum Press, 1980.

FULTON, WILLIAM. *Algebraic Curves: An Introduction to Algebraic Geometry.* Reading, MA: W.A. Benjamin, 1969.

GRIFFITHS, PHILLIP A. AND HARRIS, JOSEPH. *Principles of Algebraic Geometry.* New York, NY: John Wiley, 1978.

GRIFFITHS, PHILLIP A. *Introduction to Algebraic Curves.* Providence, RI: American Mathematical Society, 1989.

* KENDIG, KEITH. *Elementary Algebraic Geometry.* New York, NY: Springer-Verlag, 1987.

ORZECH, GRACE AND ORZECH, MORRIS. *Plane Algebraic Curves: An Introduction Via Valuations.* New York, NY: Marcel Dekker, 1981.

* REID, MILES. *Undergraduate Algebraic Geometry.* New York, NY: Cambridge University Press, 1988.

SEIDENBERG, A., ED. *Studies in Algebraic Geometry.* Washington, DC: Mathematical Association of America, 1980.

* SEIDENBERG, A. *Elements of the Theory of Algebraic Curves.* Reading, MA: Addison-Wesley, 1968.

SEMPLE, J.G. AND KNEEBONE, G.T. *Algebraic Curves.* New York, NY: Oxford University Press, 1959.

* SEMPLE, J.G. AND ROTH, L., EDS. *Introduction to Algebraic Geometry.* New York, NY: Oxford University Press, 1985.

** SHAFAREVICH, IGOR R. *Basic Algebraic Geometry.* New York, NY: Springer-Verlag, 1974.

WALKER, ROBERT J. *Algebraic Curves.* Mineola, NY: Dover, 1962; New York, NY: Springer-Verlag, 1978.

14.12 Special Topics

ARTZY, RAFAEL. *Linear Geometry.* Reading, MA: Addison-Wesley, 1974.

* DEMBOWSKI, P. *Finite Geometries.* New York, NY: Springer-Verlag, 1968.

EDMONDSON, AMY C. *A Fuller Explanation: The Synergetic Geometry of R. Buckminster Fuller.* New York, NY: Birkhäuser, 1987.

GORDON, V.O. AND SEMENTSOV-OGIEVSKII, M.A. *A Course in Descriptive Geometry.* Moscow: MIR, 1980.

HAMMER, J. *Unsolved Problems Concerning Lattice Points.* Brooklyn, NY: Pitman, 1977.

HUGHES, D.R. AND PIPER, F.C. *Design Theory,* Second Edition. New York, NY: Cambridge University Press, 1985, 1988.

** LOCKWOOD, E.H. *A Book of Curves.* New York, NY: Cambridge University Press, 1960.

LYUSTERNIK, L.A. *The Shortest Lines: Variational Problems.* Moscow: MIR, 1976, 1983.

* MORTENSON, MICHAEL E. *Geometric Modeling.* New York, NY: John Wiley, 1985.

PORTEOUS, IAN R. *Topological Geometry,* Second Edition. New York, NY: Cambridge University Press, 1981.

PUGH, ANTHONY. *An Introduction to Tensegrity.* Berkeley, CA: University of California Press, 1976.

SANTALÓ, LUIS A. *Integral Geometry and Geometric Probability.* Reading, MA: Addison-Wesley, 1976.

SCHUSTER, SEYMOUR. *Elementary Vector Geometry.* New York, NY: John Wiley, 1962.

* SNAPPER, ERNST AND TROYER, ROBERT J. *Metric Affine Geometry.* New York, NY: Academic Press, 1971; Mineola, NY: Dover, 1989.

YATES, ROBERT C. *Curves and Their Properties.* Reston, VA: National Council of Teachers of Mathematics, 1974.

15 Topology

15.1 General Topology

* ALEXANDROFF, PAUL. *Elementary Concepts of Topology.* Mineola, NY: Dover, 1961.
ARKHANGELSKIĬ, A.V. AND PONTRJAGIN, LEV S., EDS. *General Topology I: Basic Concepts and Constructions, Dimension Theory.* New York, NY: Springer-Verlag, 1990.
* BING, R.H. *Elementary Point Set Topology.* Washington, DC: Mathematical Association of America, 1960.
BOURBAKI, NICOLAS. *Elements of Mathematics: General Topology.* Reading, MA: Addison-Wesley, 1966–67; New York, NY: Springer-Verlag, 1989.
** CHINN, WILLIAM G. AND STEENROD, NORMAN E. *First Concepts of Topology.* Washington, DC: Mathematical Association of America, 1966.
DUGUNDJI, JAMES. *Topology.* Boston, MA: Allyn and Bacon, 1966.
FUKS, D.B. AND ROKHLIN, V.A. *Beginner's Course in Topology: Geometric Chapters.* New York, NY: Springer-Verlag, 1984.
GAMELIN, THEODORE W. AND GREENE, ROBERT E. *Introduction to Topology.* Philadelphia, PA: Saunders College, 1983.
GEMIGNANI, MICHAEL C. *Elementary Topology,* Second Edition. Reading, MA: Addison-Wesley, 1967; Mineola, NY: Dover, 1990.
HAUSDORFF, FELIX. *Set Theory,* Third Edition. New York, NY: Chelsea, 1957, 1978.
HUREWICZ, WITOLD AND WALLMAN, HENRY. *Dimension Theory.* Princeton, NJ: Princeton University Press, 1941.
* KAPLANSKY, IRVING. *Set Theory and Metric Spaces,* Second Edition. Needham Heights, MA: Allyn and Bacon, 1972; New York, NY: Chelsea, 1977.
* KELLEY, JOHN L. *General Topology,* New York, NY: Van Nostrand Reinhold, 1955; New York, NY: Springer-Verlag, 1975.
KURATOWSKI, K. *Topology,* 2 Vols. New York, NY: Academic Press, 1966, 1969.
*** MUNKRES, JAMES R. *Topology: A First Course.* Englewood Cliffs, NJ: Prentice Hall, 1975.
NEWMAN, M.H.A. *Elements of the Topology of Plane Sets of Points.* New York, NY: Cambridge University Press, 1964.
** OXTOBY, JOHN C. *Measure and Category: A Survey of the Analogies between Topological and Measure Spaces,* Second Edition. New York, NY: Springer-Verlag, 1980.
SIERPIŃSKI, W. *General Topology.* Toronto: University of Toronto Press, 1952.
* SIMMONS, GEORGE F. *Introduction to Topology and Modern Analysis.* New York, NY: McGraw-Hill, 1963; Melbourne, FL: Robert E. Krieger, 1983.
* STEEN, LYNN ARTHUR AND SEEBACH, J. ARTHUR, JR. *Counterexamples in Topology,* Second Edition. New York, NY: Holt, Rinehart and Winston, 1970; New York, NY: Springer-Verlag, 1978.
SUTHERLAND, W.A. *Introduction to Metric and Topological Spaces.* New York, NY: Clarendon Press, 1975.
** WILLARD, STEPHEN. *General Topology.* Reading, MA: Addison-Wesley, 1970.

15.2 Geometric Topology

* ATIYAH, MICHAEL F. *The Geometry and Physics of Knots.* New York, NY: Cambridge University Press, 1990.
BURDE, GERHARD AND ZIESCHANG, HEINER. *Knots.* Hawthorne, NY: Walter de Gruyter, 1985.
* CROWELL, RICHARD H. AND FOX, RALPH H. *Introduction to Knot Theory.* New York, NY: Springer-Verlag, 1977.
** FIRBY, P.A. AND GARDINER, C.F. *Surface Topology.* New York, NY: Halsted Press, 1982; New York, NY: Ellis Horwood, 1982.
FLEGG, H. GRAHAM. *From Geometry to Topology.* Philadelphia, PA: Crane, Russak, 1974.
* FRANCIS, GEORGE K. *A Topological Picturebook.* New York, NY: Springer-Verlag, 1987.

FREEDMAN, M.H. AND LUO, FENG. *Selected Applications of Geometry to Low-Dimensional Topology.* Providence, RI: American Mathematical Society, 1990.

* GRIFFITHS, H.B. *Surfaces,* Second Edition. New York, NY: Cambridge University Press, 1976, 1981.

KAUFFMAN, LOUIS H. *On Knots.* Princeton, NJ: Princeton University Press, 1987.

MOISE, EDWIN E. *Geometric Topology in Dimensions 2 and 3.* New York, NY: Springer-Verlag, 1977.

* MONTESINOS, JOSÉ MARIA. *Classical Tessellations and Three-Manifolds.* New York, NY: Springer-Verlag, 1987.

REIDEMEISTER, K. *Knot Theory.* Moscow, ID: BCS Associates, 1983.

** ROLFSEN, DALE. *Knots and Links,* Second Edition. Boston, MA: Publish or Perish, 1976, 1991.

ROURKE, C.P. AND SANDERSON, B.J. *Introduction to Piecewise-Linear Topology.* New York, NY: Springer-Verlag, 1972.

WALL, C.T.C. *A Geometric Introduction to Topology.* Reading, MA: Addison-Wesley, 1972.

** WEEKS, JEFFREY R. *The Shape of Space: How to Visualize Surfaces and Three-Dimensional Manifolds.* New York, NY: Marcel Dekker, 1985.

15.3 Algebraic Topology

ALEKSANDROV, P. *Combinatorial Topology,* 3 Vols. Baltimore, MD: Graylock Press, 1956–60.

** ARMSTRONG, M.A. *Basic Topology.* New York, NY: Springer-Verlag, 1983, 1990.

* ARTIN, EMIL. *Introduction to Algebraic Topology.* Columbus, OH: Charles E. Merrill, 1969.

* BLACKETT, DONALD W. *Elementary Topology: A Combinatorial and Algebraic Approach,* New York, NY: Academic Press, 1982.

GIBLIN, P.J. *Graphs, Surfaces and Homology: An Introduction to Algebraic Topology,* Second Edition. New York, NY: Halsted Press, 1977; New York, NY: Chapman and Hall, 1981.

* GRAMAIN, ANDRÉ. *Topology of Surfaces.* Moscow, ID: BCS Associates, 1984.

* GREENBERG, MARVIN JAY AND HARPER, JOHN R. *Lectures on Algebraic Topology,* Second Edition. Reading, MA: W.A. Benjamin, 1967, 1981.

*** HENLE, MICHAEL. *A Combinatorial Introduction to Topology.* New York, NY: W.H. Freeman, 1979.

JÄNICH, KLAUS. *Topology.* New York, NY: Springer-Verlag, 1984.

* KOŚNIOWSKI, CZES. *A First Course in Algebraic Topology.* New York, NY: Cambridge University Press, 1980.

LEFSCHETZ, SOLOMON. *Topology,* Second Edition. New York, NY: Chelsea, 1956.

MASSEY, WILLIAM S. *A Basic Course in Algebraic Topology.* New York, NY: Springer-Verlag, 1991.

** MASSEY, WILLIAM S. *Algebraic Topology: An Introduction.* San Diego, CA: Harcourt Brace Jovanovich, 1967; New York, NY: Springer-Verlag, 1977.

MILNOR, JOHN W. AND STASHEFF, JAMES D. *Characteristic Classes.* Princeton, NJ: Princeton University Press, 1974.

MILNOR, JOHN W. *Singular Points of Complex Hypersurfaces.* Princeton, NJ: Princeton University Press, 1968.

** MUNKRES, JAMES R. *Elements of Algebraic Topology.* Reading, MA: Addison-Wesley, 1984.

NABER, GREGORY L. *Topological Methods in Euclidean Spaces.* New York, NY: Cambridge University Press, 1980.

PONTRJAGIN, LEV S. *Foundations of Combinatorial Topology.* Baltimore, MD: Graylock Press, 1952.

PONTRJAGIN, LEV S. *Topological Groups,* Second Edition. New York, NY: Gordon and Breach, 1966.

SEIFERT, HERBERT AND THRELFALL, W. *A Textbook of Topology.* New York, NY: Academic Press, 1980.

SPANIER, EDWIN H. *Algebraic Topology.* New York, NY: McGraw-Hill, 1966.

* STILLWELL, JOHN. *Classical Topology and Combinatorial Group Theory.* New York, NY: Springer-Verlag, 1980.

* VICK, JAMES. *Homology Theory: An Introduction to Algebraic Topology.* New York, NY: Academic Press, 1973.

15.4 Differential Topology

* ALMGREN, F.J. *Plateau's Problem.* Redwood City, CA: Benjamin Cummings, 1966.

AUSLANDER, LOUIS AND MACKENZIE, ROBERT E. *Introduction to Differentiable Manifolds.* New York, NY: McGraw-Hill, 1963; Mineola, NY: Dover, 1977.

BRÖCKER, TH. AND JÄNICH, KLAUS. *Introduction to Differential Topology.* New York, NY: Cambridge University Press, 1982.

CHILLINGWORTH, D.R.J. *Differential Topology with a View to Applications.* Brooklyn, NY: Pitman, 1976.

** GUILLEMIN, VICTOR AND POLLACK, ALAN. *Differential Topology.* Englewood Cliffs, NJ: Prentice Hall, 1974.

* HIRSCH, MORRIS W. *Differential Topology.* New York, NY: Springer-Verlag, 1976.

* MILNOR, JOHN W. *Morse Theory.* Princeton, NJ: Princeton University Press, 1963.

** MILNOR, JOHN W. *Topology from the Differentiable Viewpoint.* Charlottesville, VA: University Press of Virginia, 1965.

*** SINGER, I.M. AND THORPE, JOHN A. *Lecture Notes on Elementary Topology and Geometry.* Glenview, IL: Scott Foresman, 1967; New York, NY: Springer-Verlag, 1987.

WARNER, FRANK W. *Foundations of Differentiable Manifolds and Lie Groups.* Glenview, IL: Scott Foresman, 1971; New York, NY: Springer-Verlag, 1983.

16 Vocational and Technical Mathematics

16.1 Industrial Mathematics

** AUSTIN, JACQUELINE C.; GILL, JACK C.; AND ISERN, MARGARITA. *Technical Mathematics,* Fourth Edition. Philadelphia, PA: Saunders College, 1988.

CLEAVES, CHERYL; HOBBS, MARGIE; AND DUDENHEFER, PAUL. *Introduction to Technical Mathematics.* Englewood Cliffs, NJ: Prentice Hall, 1988.

** EWEN, DALE AND NELSON, C. ROBERT. *Elementary Technical Mathematics,* Fifth Edition. Belmont, CA: Wadsworth, 1978, 1991.

KUHFITTIG, PETER K.F. *Introduction to Technical Mathematics.* Pacific Grove, CA: Brooks/Cole, 1986.

LEFFIN, WALTER W., *et al. Introduction to Technical Mathematics.* Prospect Heights, IL: Waveland, 1987.

LYNG, M.J., *et al. Applied Technical Mathematics,* Revised Edition. Boston, MA: Houghton Mifflin, 1978; Prospect Heights, IL: Waveland, 1983.

MCHALE, THOMAS J. AND WITZKE, PAUL T. *Technical Mathematics I and II.* Reading, MA: Addison-Wesley, 1988.

MOORE, CLAUDE S., *et al. Applied Math for Technicians,* Second Edition. Englewood Cliffs, NJ: Prentice Hall, 1982.

NUSTAD, HARRY L. AND WESNER, TERRY H. *Essentials of Technical Mathematics.* Dubuque, IA: William C. Brown, 1984.

PAUL, RICHARD S. AND SHAEVEL, M. LEONARD. *Essentials of Technical Mathematics,* Second Edition. Englewood Cliffs, NJ: Prentice Hall, 1982.

* SMITH, ROBERT D. *Vocational-Technical Mathematics,* Second Edition. Albany, NY: Delmar, 1983, 1991.

WALL, CHARLES R. *Basic Technical Mathematics.* San Diego, CA: Harcourt Brace Jovanovich, 1986.

** WASHINGTON, ALLYN J. AND TRIOLA, MARIO F. *Introduction to Technical Mathematics,* Fourth Edition. Redwood City, CA: Benjamin Cummings, 1988.

16.2 Mathematics for Trades

ANDERSON, J.G. *Technical Shop Mathematics,* Second Edition. New York, NY: Industrial Press, 1974, 1983.

BALL, JOHN E. *Practical Problems in Mathematics for Masons.* Albany, NY: Delmar, 1980.

* BOYCE, JOHN B., *et al. Mathematics for Technical and Vocational Students,* Eighth Edition. New York, NY: John Wiley, 1989.

BRADFORD, ROBERT. *Mathematics for Carpenters.* Albany, NY: Delmar, 1975.

CARLO, PATRICK AND MURPHY, DENNIS. *Merchandising Mathematics,* Second Edition. Albany, NY: Delmar, 1981.

CARMAN, ROBERT A. AND SAUNDERS, HAL M. *Mathematics for the Trades: A Guided Approach,* Second Edition. New York, NY: John Wiley, 1986.

* CLEAVES, CHERYL, *et al. Basic Mathematics for Trades and Technologies,* Second Edition. Englewood Cliffs, NJ: Prentice Hall, 1990.

* DEVORE, RUSSELL. *Practical Problems in Mathematics for Heating and Cooling Technicians.* Albany, NY: Delmar, 1981.

FELKER, C.A. AND BRADLEY, J.G. *Shop Mathematics,* Sixth Edition. New York, NY: McGraw-Hill, 1976, 1984.

* GARRAD, CRAWFORD G. AND HERMAN, STEPHEN L. *Practical Problems in Mathematics for Electricians,* Fourth Edition. Albany, NY: Delmar, 1987.

GOETSCH, DAVID L., *et al. Mathematics for the Automotive Trades.* Englewood Cliffs, NJ: Prentice Hall, 1988.

GOETSCH, DAVID L., *et al.* *Mathematics for the Heating, Ventilating, and Cooling Trades.* Englewood Cliffs, NJ: Prentice Hall, 1988.

GOETSCH, DAVID L., *et al.* *Mathematics for the Machine Trades.* Englewood Cliffs, NJ: Prentice Hall, 1988.

GORITZ, JOHN. *Mathematics for Welding Trades.* Englewood Cliffs, NJ: Prentice Hall, 1987.

* GUEST, RUSSELL J., *et al.* *Mathematics for Plumbers and Pipe Fitters,* Fourth Edition. Albany, NY: Delmar, 1990.

* HAINES, ROBERT G. *Math Principles for Food Service Occupations,* Second Edition. Albany, NY: Delmar, 1988.

* HENDRIX, T.G. AND LEFEVOR, C.S. *Mathematics for Auto Mechanics.* Albany, NY: Delmar, 1978.

HOFFMAN, EDWARD G. AND DAVIS, DENNIS D. *Practical Problems in Mathematics for Machinists,* Third Edition. Albany, NY: Delmar, 1988.

* HUTH, H. *Practical Problems in Mathematics for Carpenters,* Fifth Edition. Albany, NY: Delmar, 1955, 1991.

MCMACKIN, FRANK J., *et al.* *Mathematics of the Shop,* Fourth Edition. Albany, NY: Delmar, 1978.

* MOORE, GEORGE. *Practical Problems in Mathematics for Automotive Technicians,* Third Edition. Albany, NY: Delmar, 1984.

** OBERG, ERIC, *et al.* *Machinery's Handbook,* Twenty-Third Edition. New York, NY: Industrial Press, 1975, 1988.

OLIVO, C. THOMAS AND OLIVO, THOMAS P. *Basic Vocational-Technical Mathematics,* Fifth Edition. Albany, NY: Delmar, 1985.

PALMER, CLAUDE I. AND MRACHEK, LEONARD A. *Practical Mathematics,* Seventh Edition. New York, NY: McGraw-Hill, 1986.

PETERSON, JOHN C. AND DEKRYGER, WILLIAM J. *Math for the Automotive Trade,* Second Edition. Albany, NY: Delmar, 1989.

* SCHELL, FRANK R. AND MATLOCK, BILL J. *Practical Problems in Mathematics for Welders,* Third Edition. Albany, NY: Delmar, 1988.

** SMITH, ROBERT D. *Mathematics for Machine Technology,* Third Edition. Albany, NY: Delmar, 1990.

VERMEERSCH, LAVONNE F. AND SOUTHWICK, CHARLES E. *Practical Problems in Mathematics for Graphic Arts.* Albany, NY: Delmar, 1982.

* WOLFE, JOHN H. AND PHELPS, E.R. *Practical Shop Mathematics,* Fourth Edition. New York, NY: McGraw-Hill, 1958.

16.3 Health Sciences

BATASTINI, PEGGY H. AND DAVIDSON, JUDY K. *Pharmacological Calculations for Nurses: A Worktext.* Albany, NY: Delmar, 1985.

CAROLAN, MARY JANE. *Clinical Calculations for Nurses.* East Norwalk, CT: Appleton and Lange, 1990.

DANIELS, JEANNE M. AND SMITH, LORETTA M. *Clinical Calculations: A Unified Approach.* Albany, NY: Delmar, 1986.

*** HAYDEN, JEROME D. AND DAVIS, HOWARD T. *Fundamental Mathematics for Health Careers,* Second Edition. Albany, NY: Delmar, 1980, 1990.

HIGHERS, MICHAEL P. AND FORRESTER, ROBERT P. *Mathematics for the Allied Health Professions.* East Norwalk, CT: Appleton and Lange, 1987.

** HOIT, LAURA K. *The Arithmetic of Dosages and Solutions,* Seventh Edition. St. Louis, MO: Mosby and Company, 1989.

* KEE, JOYCE L. AND MARSHAL, SALLY M. *Clinical Calculations with Applications to General and Specialty Areas.* Philadelphia, PA: Saunders College, 1988.

MIDICI, GERALDINE ANN. *Drug Dosage Calculations: A Guide to Clinical Calculation.* East Norwalk, CT: Appleton and Lange, 1988.

MOORE, SUSAN G.; HOWLAND, JOSEPH W.; AND SAVAGE, KATHERINE. *Nursing Simplified: Math Logic.* Clearwater, FL: H and H Publishers, 1986.

** PICKAR, GLORIA D. *Dosage Calculations,* Third Edition. Albany, NY: Delmar, 1987, 1990.

RICE, JANE AND SKELLEY, ESTHER G. *Medications and Mathematics for the Nurse,* Sixth Edition. Albany, NY: Delmar, 1988.

* RICHARDSON, JUDITH K. AND RICHARDSON, LLOYD I. *The Mathematics of Drugs and Solutions with Clinical Applications.* St. Louis, MO: Mosby and Company, 1990.

ROBERTS, KEITH AND MICHELS, LEO. *Mathematics for Health Sciences.* Belmont, CA: Wadsworth, 1982.

WEAVER, MABEL E. AND KOEHLER, VERA J. *Programmed Mathematics of Drugs and Solutions,* Revised Edition. Philadelphia, PA: J.B. Lippincott, 1984.

WILSON, BRUCE. *Logical Nursing Mathematics.* Albany, NY: Delmar, 1987.

16.4 Data Processing

BLISS, ELIZABETH. *Data Processing Mathematics.* Englewood Cliffs, NJ: Prentice Hall, 1985.

* CALTER, PAUL. *Mathematics for Computer Technology.* Englewood Cliffs, NJ: Prentice Hall, 1986.

** CLARK, FRANK J. *Mathematics for Programming Computers,* Third Edition. Englewood Cliffs, NJ: Prentice Hall, 1988.

* DEITEL, HARVEY M. AND DEITEL, BARBARA. *Computers and Data Processing.* New York, NY: Academic Press, 1985.

KOLATIS, MARIA S. *Mathematics for Data Processing and Computing.* Reading, MA: Addison-Wesley, 1985.

* MCCULLOUGH, ROBERT N. *Mathematics for Data Processing.* Dubuque, IA: William C. Brown, 1988.

* NINESTEIN, ELEANOR H. *Introduction to Computer Mathematics.* Glenview, IL: Scott Foresman, 1987.

16.5 Electronics

BARKER, FORREST. *Problems in Technical Mathematics for Electricity/Electronics.* Redwood City, CA: Benjamin Cummings, 1976.

*** COOKE, NELSON M.; ADAMS, HERBERT F.R.; AND DELL, PETER B. *Basic Mathematics for Electronics,* Sixth Edition. New York, NY: McGraw-Hill, 1987.

GROB, BERNARD. *Mathematics for Basic Electronics,* Third Edition. New York, NY: McGraw-Hill, 1989.

PASAHOW, EDWARD. *Mathematics for Electronics.* Albany, NY: Delmar, 1984.

POWER, THOMAS C. *Electronics Mathematics.* Albany, NY: Delmar, 1985.

RADER, CARL. *Fundamentals of Electronics Mathematics.* Albany, NY: Delmar, 1985.

RICHMOND, A.E. AND HECHT, G.W. *Calculus for Electronics,* Fourth Edition. New York, NY: McGraw-Hill, 1989.

*** SINGER, B.B. AND FORSTER, H. *Basic Mathematics for Electricity and Electronics,* Sixth Edition. New York, NY: McGraw-Hill, 1976, 1989.

SULLIVAN, RICHARD L. *Modern Electronics Mathematics.* Albany, NY: Delmar, 1985.

** SULLIVAN, RICHARD L. *Practical Problems in Mathematics for Electronics Technicians,* Second Edition. Albany, NY: Delmar, 1982, 1990.

16.6 Chemical Technology

BARD, ALLEN J. *Chemical Equilibrium.* New York, NY: Harper and Row, 1966.

** GOLDFISH, DOROTHY M. *Basic Mathematics for Beginning Chemistry,* Fourth Edition. New York, NY: Macmillan, 1990.

* HAMILTON, L.F., *et al.* *Calculations of Analytic Chemistry,* Seventh Edition. New York, NY: McGraw-Hill, 1969.

MARGOLIS, EMIL J. *Chemical Principles in Calculations of Tonic Equilibria.* New York, NY: Macmillan, 1966.

NYMAN, CARL J. AND KING, G. BROOKS. *Problems for General Chemistry and Qualitative Analysis,* Fourth Edition. New York, NY: John Wiley, 1975, 1980.

PETERS, M.S. *Elementary Chemical Engineering,* Second Edition. New York, NY: McGraw-Hill, 1954, 1984.

ROBBINS, OMER, JR. *Tonic Reactions and Equilibria.* New York, NY: Macmillan, 1967.

16.7 Engineering Technology

** CALTER, PAUL. *Technical Mathematics with Calculus,* Second Edition. Englewood Cliffs, NJ: Prentice Hall, 1984, 1990.

* COOKE, NELSON M.; ADAMS, HERBERT F.R.; AND DELL, PETER B. *Basic Mathematics for Electronics with Calculus.* New York, NY: McGraw-Hill, 1989.

DAVIS, LINDA. *Technical Mathematics with Calculus.* Columbus, OH: Charles E. Merrill, 1990.

* EARLE, JAMES. *Geometry for Engineers.* Reading, MA: Addison-Wesley, 1984.

ELLIS, A.J. *Basic Algebra and Geometry for Scientists and Engineers.* New York, NY: John Wiley, 1982.

* EWEN, DALE AND TOPPER, MICHAEL A. *Mathematics for Technical Education,* Second Edition. Englewood Cliffs, NJ: Prentice Hall, 1983.

* EWEN, DALE AND TOPPER, MICHAEL A. *Technical Calculus,* Second Edition. Englewood Cliffs, NJ: Prentice Hall, 1986.

GOODSON, C.E. AND MIERTSCHIN, S.L. *Technical Mathematics with Calculus.* New York, NY: John Wiley, 1985.

** KRAMER, ARTHUR D. *Fundamentals of Technical Mathematics with Calculus,* Second Edition. New York, NY: McGraw-Hill, 1989.

* KUHFITTIG, PETER K.F. *Basic Technical Mathematics with Calculus,* Second Edition. Pacific Grove, CA: Brooks/Cole, 1984, 1989.

* PORTER, STUART R. AND ERNST, JOHN F. *Basic Technical Mathematics with Calculus.* Reading, MA: Addison-Wesley, 1985.

RICE, HAROLD S. AND KNIGHT, RAYMOND M. *Technical Mathematics with Calculus,* Third Edition. New York, NY: McGraw-Hill, 1974.

*** WASHINGTON, ALLYN J. *Basic Technical Mathematics with Calculus,* Fifth Edition. Redwood City, CA: Benjamin Cummings, 1978, 1990.

17 Business Mathematics

17.1 Basic Skills

** BERSTON, HYMAN M. AND FISHER, PAUL. *Collegiate Business Mathematics*, Fifth Edition. Homewood, IL: Richard D. Irwin, 1990.

* BITTINGER, MARVIN L. AND RUDOLPH, WILLIAM B. *Business Mathematics for College Students*. Reading, MA: Addison-Wesley, 1986.

* BOISSELLE, A.H.; FREEMAN, D.M.; AND BRENNA, L.V. *Business Mathematics Today*. New York, NY: McGraw-Hill, 1990.

LANGE, W.H.; ROUSOS, T.G.; AND MASON, R.D. *Mathematics for Business and New Consumers*. Homewood, IL: Richard D. Irwin, 1988.

** SLATER, JEFFREY. *Practical Business Math Procedures*. Homewood, IL: Richard D. Irwin, 1987.

17.2 Algebra and Finite Mathematics

* BARNETT, RAYMOND A. AND ZIEGLER, MICHAEL R. *Finite Mathematics for Management, Life, and Social Sciences*, Fourth Edition. San Francisco, CA: Dellen, 1981, 1987.

*** BARNETT, RAYMOND A.; BURKE, CHARLES J.; AND ZIEGLER, MICHAEL R. *Applied Mathematics for Business, Economics, Life Sciences, and Social Sciences*, Third Edition. San Francisco, CA: Dellen, 1989.

BEATTY, WILLIAM E. *Mathematical Relationships in Business and Economics*. Boston, MA: Prindle, Weber and Schmidt, 1970.

** BUDNICK, FRANK S. *Applied Mathematics for Business, Economics, and the Social Sciences*, Third Edition. New York, NY: McGraw-Hill, 1988.

FARLOW, STANLEY J. AND HAGGARD, GARY M. *Applied Mathematics for Business, Economics, and the Social Sciences*. New York, NY: McGraw-Hill, 1988.

* MIZRAHI, ABE AND SULLIVAN, MICHAEL. *Finite Mathematics with Applications for Business and Social Sciences*, Fifth Edition. New York, NY: John Wiley, 1983, 1988.

* NIEVERGELT, YVES. *Mathematics in Business Administration*. Homewood, IL: Richard D. Irwin, 1989.

SPENCE, LAWRENCE E.; VANDEN EYNDEN, CHARLES; AND GALLIN, DANIEL. *Applied Mathematics for the Management, Life, and Social Sciences*. Glenview, IL: Scott Foresman, 1990.

** STANCL, DONALD L. AND STANCL, MILDRED L. *Mathematics for the Management and the Life and Social Sciences*. Homewood, IL: Richard D. Irwin, 1990.

17.3 Business Statistics

* HOSSACK, I.B.; POLLARD, J.H.; AND ZEHNWIRTH, B. *Introductory Statistics with Applications in General Insurance*. New York, NY: Cambridge University Press, 1983.

MENDENHALL, WILLIAM AND MCCLAVE, JAMES T. *A Second Course in Business Statistics: Regression Analysis*, Third Edition. San Francisco, CA: Dellen, 1981, 1989.

MILLER, ROBERT B. AND WICHERN, DEAN W. *Intermediate Business Statistics*. New York, NY: Holt, Rinehart and Winston, 1977.

* NEWBOLD, PAUL. *Statistics for Business and Economics*, Second Edition. Englewood Cliffs, NJ: Prentice Hall, 1988.

*** TANUR, JUDITH M., *et al.*, EDS. *Statistics: A Guide to Business and Economics*. San Francisco, CA: Holden-Day, 1976.

17.4 Business Calculus

BERKEY, DENNIS D. *Calculus for Management, Social, and Life Sciences*, Second Edition. Philadelphia, PA: Saunders College, 1990.

** BITTINGER, MARVIN L. AND CROWN, J. CONRAD. *Mathematics and Calculus with Applications.* Reading, MA: Addison-Wesley, 1989.

BITTINGER, MARVIN L. AND MORREL, BERNARD B. *Applied Calculus,* Second Edition. Reading, MA: Addison-Wesley, 1988.

* BURGMEIER, J.W.; BOISEN, M.B., JR.; AND LARSEN, M.D. *Calculus with Applications.* New York, NY: McGraw-Hill, 1990.

* FARLOW, STANLEY J. AND HAGGARD, GARY M. *Calculus and Its Applications.* New York, NY: McGraw-Hill, 1990.

HOFFMANN, LAURENCE D. AND BRADLEY, GERALD L. *Calculus for Business, Economics, and the Social and Life Sciences,* Fourth Edition. New York, NY: McGraw-Hill, 1986, 1989.

** STANCL, DONALD L. AND STANCL, MILDRED L. *Calculus for Management and the Life and Social Sciences,* Second Edition. Homewood, IL: Richard D. Irwin, 1988, 1990.

17.5 Finance

* BROVERMAN, SAMUEL A. *Mathematics of Investment and Finance.* Winsted, CT: ACTEX Publications, 1991.

BROWN, ROBERT L. AND ZIMA, P. *Mathematics of Finance.* New York, NY: McGraw-Hill, 1983.

BUTCHER, MARJORIE V. AND NESBITT, CECIL J. *Mathematics of Compound Interest.* Ann Arbor, MI: Ulrich's Bookstore, 1971.

* CISSELL, R.; CISSELL, H.; AND FLASPOHLER, D. *Mathematics of Finance,* Fifth Edition. Boston, MA: Houghton Mifflin, 1977.

CURTIS, A.B. AND COOPER, J. *Mathematics of Accounting,* Fourth Edition. Englewood Cliffs, NJ: Prentice Hall, 1961.

HART, W.L. *Mathematics of Investment,* Fifth Edition. Lexington, MA: D.C. Heath, 1975.

** KELLISON, STEPHEN G. *The Theory of Interest,* Second Edition. Homewood, IL: Richard D. Irwin, 1970, 1990.

MCCUTCHEON, J.J. AND SCOTT, W.F. *An Introduction to the Mathematics of Finance.* London: Heinemann, 1986.

RINER, JOHN, *et al.* *Mathematics of Finance,* Fourth Edition. Englewood Cliffs, NJ: Prentice Hall, 1969.

17.6 Management

*** BOWEN, EARL K.; PRICHETT, GORDON D.; AND SABER, JOHN C. *Mathematics with Applications in Management and Economics,* Sixth Edition. Homewood, IL: Richard D. Irwin, 1972, 1987.

* COOKE, WILLIAM P. *Quantitative Methods for Management Decisions.* New York, NY: McGraw-Hill, 1985.

JOHNSON, R.H. AND WINN, P.R. *Quantitative Methods for Management.* Boston, MA: Houghton Mifflin, 1976.

* MCDONALD, T.M. *Mathematical Models for Social and Management Scientists.* Boston, MA: Houghton Mifflin, 1974.

SPRINGER, C.H., *et al.* *Mathematics for Management Sciences,* 4 Vols. Homewood, IL: Richard D. Irwin, 1965–68.

17.7 Introductory Actuarial Mathematics

BEARD, R.E.; PENTAKAINEN, T.; AND PESONEN, E. *Risk Theory.* New York, NY: Chapman and Hall, 1984.

** BOWERS, NEWTON L., JR., *et al.* *Actuarial Mathematics.* Itasca, IL: Society of Actuaries, 1984.

BUHLMAN, HANS. *Mathematical Methods in Risk Theory.* New York, NY: Springer-Verlag, 1970.

** CASUALTY ACTUARIAL SOCIETY. *Foundations of Casualty Actuarial Science.* Arlington, VA: Casualty Actuarial Society, 1990.

GERBER, HANS U. *An Introduction to Mathematical Risk Theory.* Homewood, IL: Richard D. Irwin, 1979.

GERBER, HANS U. *Life Insurance Mathematics.* New York, NY: Springer-Verlag, 1990.

* HOGG, ROBERT V. AND KLUGMAN, S. *Loss Distributions.* New York, NY: John Wiley, 1984.

JORDAN, C.W. *Life Contingencies,* Second Edition. Schaumburg, IL: Society of Actuaries, 1975.

NEILL, ALISTAIR. *Life Contingencies.* London: Heinemann, 1977.

* PARMENTER, MICHAEL M. *Theory of Interest and Life Contingencies with Pension Applications: A Problem Solving Approach.* Winsted, CT: ACTEX Publications, 1988.

STRAUB, ERWIN. *Non-Life Insurance Mathematics.* New York, NY: Springer-Verlag, 1988.

TROWBRIDGE, CHARLES L. *Fundamental Concepts of Actuarial Science.* Schaumburg, IL: Actuarial Education and Research Fund, 1989.

17.8 Advanced Actuarial Mathematics

* ANDERSON, ARTHUR W. *Pension Mathematics for Actuaries.* Needham, MA: Arthur W. Anderson, 1985.

ANDREWS, GEORGE E. AND BEEKMAN, JOHN A. *Actuarial Projections for the Old Age, Survivor, and Disability Income Program of Social Security in the United States of America.* Schaumburg, IL: Society of Actuaries, 1987.

BATTEN, R.W. *Mortality Table Construction.* Englewood Cliffs, NJ: Prentice Hall, 1978.

* BERIN, BARNET N. *The Fundamentals of Pension Mathematics.* Schaumburg, IL: Society of Actuaries, 1989.

BROWN, ROBERT L. *Introduction to Property Casualty Ratemaking and Loss Reserving.* Winsted, CT: ACTEX Publications, 1991.

CHIANG, CHIN LONG. *The Life Table and Its Applications.* Melbourne, FL: Robert E. Krieger, 1984.

GERSCHENSON, H. *Measurement of Mortality.* Schaumburg, IL: Society of Actuaries, 1961.

HUSTEAD, E.C. *100 Years of Mortality.* Schaumburg, IL: Society of Actuaries, 1989.

LONDON, DICK. *Graduation: The Revision of Estimates.* Winsted, CT: ACTEX Publications, 1985.

LONDON, DICK. *Survival Models and their Estimation,* Second Edition. Winsted, CT: ACTEX Publications, 1988.

MOORHEAD, E.J. *Our Yesterdays: The History of the Actuarial Profession in North America, 1805–1979.* Schaumburg, IL: Society of Actuaries, 1989.

* PANJER, HARRY E., ED. *Actuarial Mathematics.* Providence, RI: American Mathematical Society, 1985.

18 Numerical Analysis

18.1 Introductory Texts

* ATKINSON, KENDALL E. *Elementary Numerical Analysis.* New York, NY: John Wiley, 1978, 1985.

*** BURDEN, RICHARD L. AND FAIRES, J. DOUGLAS. *Numerical Analysis,* Fourth Edition. Boston, MA: PWS-Kent, 1989.

CHENEY, ELLIOT W. AND KINCAID, DAVID R. *Numerical Mathematics and Computing,* Second Edition. Pacific Grove, CA: Brooks/Cole, 1980, 1985.

* CONTE, SAMUEL D. AND DE BOOR, CARL. *Elementary Numerical Analysis: An Algorithmic Approach,* Third Edition. New York, NY: McGraw-Hill, 1972, 1980.

** DAVIS, PHILIP F. AND RABINOWITZ, PHILIP. *Methods of Numerical Integration,* Second Edition. New York, NY: Academic Press, 1975, 1984.

DORN, WILLIAM S. AND MCCRACKEN, DANIEL D. *Numerical Methods with FORTRAN IV Case Studies.* New York, NY: John Wiley, 1972.

* GERALD, CURTIS F. AND WHEATLEY, PATRICK O. *Applied Numerical Analysis,* Fourth Edition. Reading, MA: Addison-Wesley, 1978, 1989.

* HAMMING, RICHARD W. *Introduction to Applied Numerical Analysis.* New York, NY: Hemisphere, 1989.

JOHNSON, LEE W. AND RIESS, R. DEAN. *Numerical Analysis,* Second Edition. Reading, MA: Addison-Wesley, 1977, 1982.

*** KINCAID, DAVID R. AND CHENEY, ELLIOT W. *Numerical Analysis: Mathematics of Scientific Computing.* Pacific Grove, CA: Brooks/Cole, 1991.

MARON, MELVIN J. AND LOPEZ, ROBERT J. *Numerical Analysis: A Practical Approach,* Third Edition. New York, NY: Macmillan, 1982; Belmont, CA: Wadsworth, 1991.

* MATHEWS, JOHN H. *Numerical Methods for Computer Science, Engineering, and Mathematics.* Englewood Cliffs, NJ: Prentice Hall, 1987.

MORRIS, J. LL. *Computational Methods in Elementary Numerical Analysis.* New York, NY: John Wiley, 1983.

RALSTON, ANTHONY AND RABINOWITZ, PHILIP. *A First Course in Numerical Analysis,* Second Edition. New York, NY: McGraw-Hill, 1978.

** STOER, J. AND BULIRSCH, R. *Introduction to Numerical Analysis.* New York, NY: Springer-Verlag, 1980.

18.2 Advanced Surveys

* CARNAHAN, BRICE; LUTHER, H.A.; AND WILKES, JAMES O. *Applied Numerical Methods.* New York, NY: John Wiley, 1969.

CHURCHHOUSE, ROBERT F., ED. *Numerical Methods.* Handbook of Applicable Mathematics, Volume III. New York, NY: John Wiley, 1981.

DAHLQUIST, GERMUND AND ÅKE BJÖRCK. *Numerical Methods.* Englewood Cliffs, NJ: Prentice Hall, 1974.

** GOLUB, GENE H., ED. *Studies in Numerical Analysis.* Washington, DC: Mathematical Association of America, 1984.

** GREENSPAN, DONALD AND CASULLI, VINCENZO. *Numerical Analysis for Applied Mathematics, Science, and Engineering.* Reading, MA: Addison-Wesley, 1988.

JACQUES, IAN AND JUDD, COLIN. *Numerical Analysis.* New York, NY: Chapman and Hall, 1987.

JAIN, M.K.; IYENGAR, S.R.K.; AND JAIN, R.K. *Numerical Methods for Scientific and Engineering Computation.* New York, NY: Halsted Press, 1985.

* ORTEGA, JAMES M. *Numerical Analysis: A Second Course.* New York, NY: Academic Press, 1972; Philadelphia, PA: Society for Industrial and Applied Mathematics, 1990.

TODD, JOHN. *Basic Numerical Mathematics,* 2 Vols. New York, NY: Birkhäuser, 1977, 1979.

TRAUB, JOSEPH F. *Iterative Methods for the Solution of Equations.* New York, NY: Chelsea, 1982.

VANDERGRAFT, JAMES S. *Introduction to Numerical Computations.* New York, NY: Academic Press, 1978.

YOUNG, DAVID M. AND GREGORY, ROBERT T. *A Survey of Numerical Mathematics,* 2 Vols. Reading, MA: Addison-Wesley, 1972, 1973.

18.3 Differential Equations

AXELSSON, O. AND BARKER, V.A. *Finite Element Solution of Boundary Value Problems: Theory and Computation.* New York, NY: Academic Press, 1984.

* BATHE, KLAUS-JÜRGEN AND WILSON, EDWARD L. *Numerical Methods in Finite Element Analysis.* Englewood Cliffs, NJ: Prentice Hall, 1976.

CIARLET, PHILIPPE G. *The Finite Element Method for Elliptic Problems.* Amsterdam: North-Holland, 1978.

HALL, G. AND WATT, J.M., EDS. *Modern Numerical Methods for Ordinary Differential Equations.* New York, NY: Clarendon Press, 1976.

HENRICI, PETER. *Discrete Variable Methods in Ordinary Differential Equations.* New York, NY: John Wiley, 1962.

JAIN, M.K. *Numerical Solution of Differential Equations.* New York, NY: Halsted Press, 1979.

* KELLER, H.B. *Numerical Methods for Two-Point Boundary-Value Problems.* New York, NY: John Wiley, 1968.

* LAPIDUS, LEON AND PINDER, GEORGE F. *Numerical Solution of Partial Differential Equations in Science and Engineering.* New York, NY: John Wiley, 1982.

MEIS, THEODOR AND MARCOWITZ, ULRICH. *Numerical Solution of Partial Differential Equations.* New York, NY: Springer-Verlag, 1981.

PARKER, THOMAS S. AND CHUA, LEON O. *Practical Numerical Algorithms for Chaotic Systems.* New York, NY: Springer-Verlag, 1989.

* QUINNEY, DOUGLAS. *An Introduction to the Numerical Solution of Differential Equations.* New York, NY: John Wiley, 1985.

* SEWELL, GRANVILLE. *The Numerical Solution of Ordinary and Partial Differential Equations.* New York, NY: Academic Press, 1988.

* SMITH, G.D. *Numerical Solution of Partial Differential Equations: Finite Difference Methods,* Third Edition. New York, NY: Oxford University Press, 1969; New York, NY: Clarendon Press, 1978, 1985.

* SOD, GARY A. *Numerical Methods in Fluid Dynamics: Initial and Initial Boundary-Value Problems.* New York, NY: Cambridge University Press, 1985.

TWIZELL, E.H. *Computational Methods for Partial Differential Equations.* New York, NY: Halsted Press, 1984.

VEMURI, V. AND KARPLUS, WALTER J. *Digital Computer Treatment of Partial Differential Equations.* Englewood Cliffs, NJ: Prentice Hall, 1981.

18.4 Numerical Linear Algebra

** CIARLET, PHILIPPE G. *Numerical Linear Algebra.* New York, NY: Cambridge University Press, 1989.

** COLEMAN, THOMAS F. AND VAN LOAN, CHARLES F. *Handbook for Matrix Computations.* Philadelphia, PA: Society for Industrial and Applied Mathematics, 1988.

FORSYTHE, GEORGE E. AND MOLER, CLEVE B. *Computer Solution of Linear Algebraic Systems.* Englewood Cliffs, NJ: Prentice Hall, 1967.

*** GOLUB, GENE H. AND VAN LOAN, CHARLES F. *Matrix Computations,* Second Edition. Baltimore, MD: Johns Hopkins University Press, 1983, 1989.

HAGER, WILLIAM. *Applied Numerical Linear Algebra.* Englewood Cliffs, NJ: Prentice Hall, 1988.

* HILL, DAVID R. AND MOLER, CLEVE B. *Experiments in Computational Matrix Algebra.* Cambridge, MA: Random House, 1987.

JENNINGS, ALAN. *Matrix Computation for Engineers and Scientists.* New York, NY: John Wiley, 1977.

* MAGID, ANDY R. *Applied Matrix Models: A Second Course in Linear Algebra with Computer Applications.* New York, NY: John Wiley, 1985.

PISSANETZKY, SERGIO. *Sparse Matrix Technology.* New York, NY: Academic Press, 1984.

* RICE, JOHN R. *Matrix Computations and Mathematical Software.* New York, NY: McGraw-Hill, 1981.

STEWART, G.W. *Introduction to Matrix Computation.* New York, NY: Academic Press, 1973.

VARGA, RICHARD S. *Matrix Iterative Analysis.* Englewood Cliffs, NJ: Prentice Hall, 1962.

*** WILKINSON, JAMES H. *The Algebraic Eigenvalue Problem.* New York, NY: Oxford University Press, 1965, 1988.

YOUNG, DAVID M. *Iterative Solution of Large Linear Systems.* New York, NY: Academic Press, 1971.

18.5 Approximation Theory

*** CHENEY, ELLIOT W. *Introduction to Approximation Theory,* Second Edition. New York, NY: Chelsea, 1966, 1982.

** DAVIS, PHILIP J. *Interpolation and Approximation.* Mineola, NY: Dover, 1975.

** DE BOOR, CARL. *A Practical Guide to Splines.* New York, NY: Springer-Verlag, 1978.

DE BRUIJN, N.G. *Asymptotic Methods in Analysis.* Mineola, NY: Dover, 1981.

FEINERMAN, ROBERT P. AND NEWMAN, DONALD J. *Polynomial Approximation.* Baltimore, MD: Williams and Wilkins, 1974.

HESTENES, MAGNUS R. *Conjugate Direction Methods in Optimization.* New York, NY: Springer-Verlag, 1980.

* LORENTZ, G.G. *Approximation of Functions,* Second Edition. New York, NY: Chelsea, 1986.

PRENTER, P.M. *Splines and Variational Methods.* New York, NY: John Wiley, 1975.

RIVLIN, THEODORE J. *An Introduction to the Approximation of Functions.* Mineola, NY: Dover, 1981.

* WAHBA, GRACE. *Spline Models for Observational Data.* Philadelphia, PA: Society for Industrial and Applied Mathematics, 1990.

* WATSON, G.A. *Approximation Theory and Numerical Methods.* New York, NY: John Wiley, 1980.

18.6 Computer Methods

* FORSYTHE, GEORGE E.; MALCOLM, MICHAEL A.; AND MOLER, CLEVE B. *Computer Methods for Mathematical Computations.* Englewood Cliffs, NJ: Prentice Hall, 1977.

* FRÖBERG, CARL-ERIK. *Numerical Mathematics: Theory and Computer Applications.* Redwood City, CA: Benjamin Cummings, 1985.

** GRANDINE, THOMAS A. *The Numerical Methods Programming Projects Book.* New York, NY: Oxford University Press, 1990.

* JAMES, M.L.; SMITH, G.M.; AND WOLFORD, J.C. *Applied Numerical Methods for Digital Computation,* Third Edition. New York, NY: Harper and Row, 1985.

JOHNSTON, R.L. *Numerical Methods: A Software Approach.* New York, NY: John Wiley, 1982.

KAHANER, DAVID; MOLER, CLEVE B.; AND NASH, STEPHEN. *Numerical Methods and Software.* Englewood Cliffs, NJ: Prentice Hall, 1989.

KOPCHENOVA, N.V. AND MARON, I.A. *Computational Mathematics: Worked Examples and Problems with Elements of Theory.* Moscow: MIR, 1981.

LASTMAN, GARY J. AND SINHA, NARESH K. *Microcomputer-Based Numerical Methods for Science and Engineering.* New York, NY: W.B. Saunders, 1989.

* MOHAMED, J.L. AND WALSH, J.E., EDS. *Numerical Algorithms.* New York, NY: Clarendon Press, 1986.

*** PRESS, WILLIAM H., *et al. Numerical Recipes: The Art of Scientific Computing.* New York, NY: Cambridge University Press, 1986.

SCHENDEL, U. *Introduction to Numerical Methods for Parallel Computers.* New York, NY: Halsted Press, 1984.

18.7 Special Topics

BAKER, CHRISTOPHER T.H. *The Numerical Treatment of Integral Equations.* New York, NY: Clarendon Press, 1977.

BRIGHAM, E. ORAN. *The Fast Fourier Transform.* Englewood Cliffs, NJ: Prentice Hall, 1974.

* DELVES, L.M. AND MOHAMED, J.L. *Computational Methods for Integral Equations.* New York, NY: Cambridge University Press, 1985.

IRONS, BRUCE AND SHRIVE, NIGEL G. *Finite Element Primer.* New York, NY: Halsted Press, 1983.

ZIENKIEWICZ, O.C. *The Finite Element Method.* New York, NY: McGraw-Hill, 1977.

19 Modeling and Operations Research

19.1 General

* DAELLENBACH, HANS G. AND GEORGE, JOHN A. *Introduction to Operations Research Techniques.* Boston, MA: Allyn and Bacon, 1978.

* DANTZIG, GEORGE B. AND EAVES, B.C., EDS. *Studies in Optimization.* Washington, DC: Mathematical Association of America, 1974.

FOULDS, L.R. *Combinatorial Optimization for Undergraduates.* New York, NY: Springer-Verlag, 1984.

FRENCH, SIMON. *Readings in Decision Analysis.* New York, NY: Chapman and Hall, 1989.

* GASS, SAUL I. *Decision Making, Models, and Algorithms: A First Course.* New York, NY: John Wiley, 1985.

GRIBIK, PAUL R. AND KORTANEK, KENNETH O. *Extremal Methods of Operations Research.* New York, NY: Marcel Dekker, 1985.

*** HILLIER, FREDERICK S. AND LIEBERMAN, GERALD J. *Introduction to Operations Research,* Fifth Edition. Oakland, CA: Holden-Day, 1974; New York, NY: McGraw-Hill, 1990.

MARLOW, W.H. *Mathematics for Operations Research.* New York, NY: John Wiley, 1978.

SCHMIDT, J. WILLIAM AND DAVIS, ROBERT P. *Foundations of Analysis in Operations Research.* New York, NY: Academic Press, 1981.

* SCHMIDT, J. WILLIAM. *Mathematical Foundations for Management Science and Systems Analysis.* New York, NY: Academic Press, 1974.

TRUEMAN, RICHARD E. *An Introduction to Quantitative Methods for Decision Making,* Second Edition. New York, NY: Holt, Rinehart and Winston, 1977.

*** WAGNER, HARVEY M. *Principles of Operations Research with Applications to Managerial Decisions,* Second Edition. Englewood Cliffs, NJ: Prentice Hall, 1969, 1975.

** WINSTON, WAYNE L. *Operations Research: Applications and Algorithms,* Second Edition. Boston, MA: Duxbury Press, 1987; Boston, MA: PWS-Kent, 1991.

19.2 Mathematical Modeling

* ANDREWS, J.G. AND MCLONE, R.R., EDS. *Mathematical Modelling.* Woburn, MA: Butterworth, 1976.

* BELTRAMI, EDWARD J. *Mathematics for Dynamic Modeling.* New York, NY: Academic Press, 1987.

* BOYCE, WILLIAM E., ED. *Case Studies in Mathematical Modeling.* Brooklyn, NY: Pitman, 1981.

** BRAUN, MARTIN; COLEMAN, COURTNEY S.; AND DREW, DONALD A., EDS. *Differential Equation Models.* Modules in Applied Mathematics, Vol. 1. New York, NY: Springer-Verlag, 1983.

BURGHES, DAVID N.; HUNTLEY, IAN; AND MCDONALD, JOHN. *Applying Mathematics: A Course in Mathematical Modelling.* New York, NY: Ellis Horwood, 1982.

CROSS, MARK AND MOSCARDINI, A.O. *Learning the Art of Mathematical Modelling.* New York, NY: Halsted Press, 1985.

GASS, SAUL I., ED. *Operations Research: Mathematics and Models.* Providence, RI: American Mathematical Society, 1981.

* GIORDANO, FRANK R. AND WEIR, MAURICE D. *A First Course in Mathematical Modeling.* Pacific Grove, CA: Brooks/Cole, 1985.

** HABERMAN, RICHARD. *Mathematical Models, Mechanical Vibrations, Population Dynamics, and Traffic Flow: An Introduction to Applied Mathematics.* Englewood Cliffs, NJ: Prentice Hall, 1977.

* HUA, LOO-KENG AND WANG, YUAN. *Popularizing Mathematical Methods in the People's Republic of China: Some Personal Experiences.* New York, NY: Birkhäuser, 1989.

JAMES, D.J.G. AND MCDONALD, JOHN, EDS. *Case Studies in Mathematical Modelling.* New York, NY: Halsted Press, 1981.

KAPUR, J.N. *Mathematical Modelling.* New York, NY: John Wiley, 1988.

** KLAMKIN, MURRAY S. *Mathematical Modelling: Classroom Notes in Applied Mathematics.* Philadelphia, PA: Society for Industrial and Applied Mathematics, 1987.

** LUCAS, WILLIAM F.; ROBERTS, FRED S.; AND THRALL, ROBERT M., EDS. *Discrete and System Models.* Modules in Applied Mathematics, Vol. 3. New York, NY: Springer-Verlag, 1983.

* MELZAK, Z.A. *Bypasses: A Simple Approach to Complexity.* New York, NY: John Wiley, 1983.

MELZAK, Z.A. *Mathematical Ideas, Modeling, and Applications.* New York, NY: John Wiley, 1976.

MESTERTON-GIBBONS, MICHAEL. *A Concrete Approach to Mathematical Modelling.* Reading, MA: Addison-Wesley, 1989.

* MEYER, WALTER J. *Concepts of Mathematical Modeling.* New York, NY: McGraw-Hill, 1984.

MOLLOY, MICHAEL K. *Fundamentals of Performance Modeling.* New York, NY: Macmillan, 1989.

NICOLAS, GREGOIER AND PRIGOGINE, ILYA. *Exploring Complexity: An Introduction.* New York, NY: W.H. Freeman, 1989.

** POSTON, TIM AND STEWART, IAN. *Catastrophe Theory and its Applications.* Brooklyn, NY: Pitman, 1978.

RICHARDSON, JACQUES, ED. *Models of Reality: Shaping Thought and Action.* Mt. Airy, MD: Lomond, 1984.

* ROBERTS, FRED S. *Discrete Mathematical Models with Applications to Social, Biological, and Environmental Problems.* Englewood Cliffs, NJ: Prentice Hall, 1976.

SAATY, THOMAS L. AND ALEXANDER, JOYCE M. *Thinking With Models: Mathematical Models in the Physical, Biological, and Social Sciences.* Elmsford, NY: Pergamon Press, 1981.

STARFIELD, ANTHONY M.; SMITH, KARL A.; AND BLELOCH, ANDREW L. *How to Model It: Problem Solving for the Computer Age.* New York, NY: McGraw-Hill, 1990.

* THOMPSON, JAMES R. *Empirical Model Building.* New York, NY: John Wiley, 1989.

* WAN, FREDERIC Y.M. *Mathematical Models and Their Analysis.* New York, NY: Harper and Row, 1989.

WILLIAMS, H.P. *Model Building in Mathematical Programming,* Second Edition. New York, NY: John Wiley, 1978, 1985.

** WOODCOCK, ALEXANDER AND DAVIS, MONTE. *Catastrophe Theory.* New York, NY: E.P. Dutton, 1978.

ZEEMAN, E.C. *Catastrophe Theory: Selected Papers, 1972–1977.* Reading, MA: Addison-Wesley, 1977.

19.3 Stochastic Modeling and Dynamic Programming

ATHREYA, A.K.B. AND NEY, P.E. *Branching Processes.* New York, NY: Springer-Verlag, 1972.

BAILEY, NORMAN T.J. *Mathematical Theory of Epidemics.* London: Charles Griffin, 1957.

* BARLOW, RICHARD AND PROSCHAN, FRANK. *Statistical Theory of Reliability and Life Testing.* New York, NY: Holt, Rinehart and Winston, 1975.

DENARDO, ERIC V. *Dynamic Programming: Models and Applications.* Englewood Cliffs, NJ: Prentice Hall, 1982.

DREYFUS, STUART E. AND LAW, AVERILL M. *The Art and Theory of Dynamic Programming.* New York, NY: Academic Press, 1977.

FRENCH, SIMON. *Sequencing and Scheduling: An Introduction to the Methodologies of the Job-Shop.* New York, NY: Ellis Horwood, 1982.

** GAZIS, DENOS C., ED. *Traffic Science.* New York, NY: John Wiley, 1974.

GETIS, ARTHUR AND BOOTS, BARRY. *Models of Spatial Processes: An Approach to the Study of Point, Line, and Area Patterns.* New York, NY: Cambridge University Press, 1978.

GITTINS, J.C. *Multi-Armed Bandit Allocation Indices.* New York, NY: John Wiley, 1989.

HAIGHT, F.A. *Mathematical Theories of Traffic Flow.* New York, NY: Academic Press, 1963.

* HEYMAN, D.P. AND SOBEL, M.J. *Handbook in Operations Research and Management Science.* Amsterdam: North-Holland, 1989.

*** HEYMAN, D.P. AND SOBEL, M.J. *Stochastic Models in Operations Research,* 2 Vols. New York, NY: McGraw-Hill, 1982, 1984.

PRABHU, N.U. *Stochastic Storage Processes: Queues, Insurance Risk, and Dams.* New York, NY: Springer-Verlag, 1980.

** ROSS, SHELDON M. *Introduction to Stochastic Dynamic Programming.* New York, NY: Academic Press, 1983.

SAUER, C.H. AND CHANDY, K. *Computer Systems Performance Modeling.* Englewood Cliffs, NJ: Prentice Hall, 1981.

SEAL, H.L. *Survival Probabilities: The Goal of Risk Theory.* New York, NY: John Wiley, 1978.

SYSKI, R. *Introduction to Congestion Theory in Telephone Systems.* Amsterdam: North-Holland, 1986.

WHITTLE, PETER. *Optimization Over Time: Programming and Stochastic Control,* 2 Vols. New York, NY: John Wiley, 1983.

19.4 Game Theory

** AUMANN, ROBERT J. AND HART, SERGIU. *Handbook of Game Theory with Applications to Economics,* Amsterdam: North-Holland, 1991.

BLACKWELL, DAVID AND GIRSHICK, M.A. *Theory of Games and Statistical Decisions.* New York, NY: John Wiley, 1954; Mineola, NY: Dover, 1979.

BRAMS, STEVEN J. AND KILGOUR, D. MARC. *Game Theory and National Security.* Cambridge, MA: Basil Blackwell, 1988.

* DAVIS, MORTON D. *Game Theory: A Nontechnical Introduction.* New York, NY: Basic Books, 1970.

DRESHER, MELVIN. *The Mathematics of Games of Strategy: Theory and Applications.* Mineola, NY: Dover, 1981.

HÁJEK, OTOMAR. *Pursuit Games: An Introduction to the Theory and Applications of Differential Games of Pursuit and Evasion.* New York, NY: Academic Press, 1975.

JONES, A.J. *Game Theory: Mathematical Models of Conflict.* New York, NY: Halsted Press, 1980.

* LUCAS, WILLIAM F., ED. *Game Theory and its Applications.* Providence, RI: American Mathematical Society, 1981.

*** LUCE, ROBERT DUNCAN AND RAIFFA, HOWARD. *Games and Decisions.* New York, NY: John Wiley, 1957.

* MCDONALD, JOHN. *The Game of Business.* New York, NY: Doubleday, 1975.

** MYERSON, ROGER B. *Game Theory: Analysis of Conflict.* Cambridge, MA: Harvard University Press, 1991.

OWEN, GUILLERMO. *Game Theory,* Second Edition. New York, NY: W.B. Saunders, 1968; New York, NY: Academic Press, 1982.

RUCKLE, W.H. *Geometric Games and Their Applications.* Brooklyn, NY: Pitman, 1983.

* SHUBIK, MARTIN. *Game Theory in the Social Sciences,* 2 Vols. Cambridge, MA: MIT Press, 1982, 1984.

STONE, LAWRENCE D. *Theory of Optimal Search.* New York, NY: Academic Press, 1975.

THOMAS, L.C. *Games: Theory and Applications.* New York, NY: Halsted Press, 1984.

*** VON NEUMANN, JOHN AND MORGENSTERN, OSKAR. *Theory of Games and Economic Behavior.* Princeton, NJ: Princeton University Press, 1980.

WANG, JIANHUA. *The Theory of Games.* New York, NY: Clarendon Press, 1988.

** WILLIAMS, JOHN D. *The Compleat Strategyst: Being a Primer on the Theory of Games of Strategy.* New York, NY: McGraw-Hill, 1954; Mineola, NY: Dover, 1986.

19.5 Linear Programming

BRICKMAN, LOUIS. *Mathematical Introduction to Linear Programming and Game Theory.* New York, NY: Springer-Verlag, 1989.

BUCHANAN, J.T. *Discrete and Dynamic Decision Analysis.* New York, NY: John Wiley, 1982.

*** CHVÁTAL, VASEK. *Linear Programming.* New York, NY: W.H. Freeman, 1983.

COLLATZ, L. AND WETTERLING, W. *Optimization Problems.* New York, NY: Springer-Verlag, 1975.

*** DANTZIG, GEORGE B. *Linear Programming and Extensions.* Princeton, NJ: Princeton University Press, 1963.

** GASS, SAUL I. *Linear Programming: Methods and Applications,* Fifth Edition. New York, NY: McGraw-Hill, 1969, 1985.

HARTLEY, ROGER. *Linear and Nonlinear Programming: An Introduction to Linear Methods in Mathematical Programming.* New York, NY: Halsted Press, 1985.

KAPLAN, EDWARD L. *Mathematical Programming and Games.* New York, NY: John Wiley, 1982.

* KOLMAN, BERNARD AND BECK, ROBERT E. *Elementary Linear Programming with Applications,* New York, NY: Academic Press, 1980.

LUENBERGER, D. *Linear and Nonlinear Programming,* Second Edition. Reading, MA: Addison-Wesley, 1984.

MURTY, KATTA G. *Linear and Combinatorial Programming.* New York, NY: John Wiley, 1976; Melbourne, FL: Robert E. Krieger, 1985.

NAZARETH, J.L. *Computer Solution of Linear Programs.* New York, NY: Oxford University Press, 1987.

* SOLODOVNIKOV, A.S. *Systems of Linear Inequalities.* Chicago, IL: University of Chicago Press, 1980.

* SOLOW, DANIEL. *Linear Programming: An Introduction to Finite Improvement Algorithms.* New York, NY: Elsevier Science, 1984.

STANCU-MINASIAN, I.M. *Stochastic Programming with Multiple Objective Functions.* Norwell, MA: D. Reidel, 1984.

STRAYER, JAMES K. *Linear Programming and Its Applications.* New York, NY: Springer-Verlag, 1989.

* THIE, PAUL R. *An Introduction to Linear Programming and Game Theory,* Second Edition. New York, NY: John Wiley, 1979, 1988.

WALSH, G.R. *An Introduction to Linear Programming,* Second Edition. New York, NY: John Wiley, 1985.

19.6 Nonlinear Programming

* COHON, JARED L. *Multiobjective Programming and Planning.* New York, NY: Academic Press, 1978.

GILL, PHILIP E.; MURRAY, WALTER; AND WRIGHT, MARGARET H. *Practical Optimization.* New York, NY: Academic Press, 1981.

MCCORMICK, GARTH P. *Nonlinear Programming: Theory, Algorithms, and Applications.* New York, NY: John Wiley, 1983.

* PERESSINI, ANTHONY L.; SULLIVAN, FRANCIS E.; AND UHL, J.J., JR. *The Mathematics of Nonlinear Programming.* New York, NY: Springer-Verlag, 1988.

* SAATY, THOMAS L. AND BRAM, JOSEPH. *Nonlinear Mathematics.* Mineola, NY: Dover, 1981.

SHAPIRO, ROY D. *Optimization Models for Planning and Allocation: Text and Cases in Mathematical Programming.* New York, NY: John Wiley, 1984.

** SIMMONS, DONALD M. *Nonlinear Programming for Operations Research.* Englewood Cliffs, NJ: Prentice Hall, 1975.

WISMER, DAVID A. AND CHATTERGY, R. *Introduction to Nonlinear Optimization: A Problem Solving Approach.* Amsterdam: North-Holland, 1978.

19.7 Integer Programming

* FORD, L.R., JR. AND FULKERSON, D.R. *Flows in Networks.* Princeton, NJ: Princeton University Press, 1962.

GARFINKEL, ROBERT S. AND NEMHAUSER, GEORGE L. *Integer Programming.* New York, NY: John Wiley, 1972.

** LAWLER, EUGENE L., *et al.,* EDS. *The Traveling Salesman Problem: A Guided Tour of Combinatorial Optimization.* New York, NY: John Wiley, 1985.

LAWLER, EUGENE L. *Combinatorial Optimization: Networks and Matroids.* New York, NY: Holt, Rinehart and Winston, 1976.

*** NEMHAUSER, GEORGE L. AND WOLSEY, LAURENCE A. *Integer and Combinatorial Optimization.* New York, NY: John Wiley, 1988.

* PAPADIMITRIOU, CHRISTOS H. AND STEIGLITZ, KENNETH. *Combinatorial Optimization: Algorithms and Complexity.* Englewood Cliffs, NJ: Prentice Hall, 1982.

PARKER, R. GARY AND RARDIN, RONALD L. *Discrete Optimization.* New York, NY: Academic Press, 1988.

SAATY, THOMAS L. *Optimization in Integers and Related Extremal Problems.* New York, NY: McGraw-Hill, 1970.

* SCHRIJVER, ALEXANDER. *Theory of Linear and Integer Programming.* New York, NY: John Wiley, 1986.

TAHA, HAMDY A. *Integer Programming: Theory, Applications, and Computations.* New York, NY: Academic Press, 1975.

19.8 Queueing Theory

ALLEN, ARNOLD O. *Probability, Statistics, and Queueing Theory with Computer Science Applications,* Second Edition. New York, NY: Academic Press, 1977, 1990.

BOROVKOV, A.A. *Asymptotic Methods in Queueing Theory.* New York, NY: John Wiley, 1984.

BOROVKOV, A.A. *Stochastic Processes in Queueing Theory.* New York, NY: Springer-Verlag, 1976.

COHEN, J.W. *The Single Server Queue,* Second Edition. Amsterdam: North-Holland, 1982.

* COOPER, R.B. *Introduction to Queueing Theory,* Second Edition. Amsterdam: North-Holland, 1981.

FRANKEN, P., *et al. Queues and Point Processes.* New York, NY: John Wiley, 1982.

GELENBE, E. AND PUJOLLE, G. *Introduction to Queueing Networks.* New York, NY: John Wiley, 1987.

*** GROSS, DONALD AND HARRIS, CARL M. *Fundamentals of Queueing Theory,* Second Edition. New York, NY: John Wiley, 1974, 1985.

KALASHNIKOV, V.V. AND RACHEV, S.T. *Mathematical Models for the Construction of Queueing Models.* Belmont, CA: Wadsworth, 1990.

** KLEINROCK, L. *Queueing Systems,* 2 Vols. New York, NY: John Wiley, 1976.

NEUTS, MARCEL. *Structured Stochastic Matrices of M/G/1 Type and Their Applications.* New York, NY: Marcel Dekker, 1989.

WALRAND, J. *An Introduction to Queueing Networks.* Englewood Cliffs, NJ: Prentice Hall, 1988.

* WOLFF, R.W. *Stochastic Modelling and the Theory of Queues.* Englewood Cliffs, NJ: Prentice Hall, 1989.

19.9 Simulation

BANKS, JERRY. *Discrete-Event System Simulation.* Englewood Cliffs, NJ: Prentice Hall, 1984.

** BRATLEY, PAUL; FOX, BENNETT L.; AND SCHRAGE, LINUS E. *A Guide to Simulation.* New York, NY: Springer-Verlag, 1983.

DEVROYE, L. *Non-Uniform Random Variate Generation.* New York, NY: Springer-Verlag, 1986.

** FISHMAN, G.S. *Principles of Discrete Event Simulation.* New York, NY: John Wiley, 1978.

GRAYBEAL, WAYNE J. AND POOCH, UDO W. *Simulation: Principles and Methods.* Cambridge: Winthrop, 1980.

LAW, AVERILL M. AND KELTON, W.D. *Simulation Modeling and Analysis.* New York, NY: McGraw-Hill, 1982.

MORGAN, BYRON J.T. *Elements of Simulation.* New York, NY: Chapman and Hall, 1984.

ORD-SMITH, R.J. AND STEPHENSON, J. *Computer Simulation of Continuous Systems.* New York, NY: Cambridge University Press, 1975.

* RIPLEY, BRIAN D. *Stochastic Simulation.* New York, NY: John Wiley, 1987.

RUBINSTEIN, REUVEN Y. *Monte Carlo Optimization, Simulation, and Sensitivity of Queueing Networks.* New York, NY: John Wiley, 1986.

RUBINSTEIN, REUVEN Y. *Simulation and the Monte Carlo Method.* New York, NY: John Wiley, 1981.

* VAN DAAL, J. AND VAN DOELAND, F. *On the Height of a Railway Bridge.* Beaverton, OR: Rotterdam University Press, 1974.

19.10 Control Theory

** BARNETT, STEPHEN AND CAMERON, R.G. *Introduction to Mathematical Control Theory,* Second Edition. New York, NY: Oxford University Press, 1975; New York, NY: Clarendon Press, 1985.

BARNETT, STEPHEN. *Polynomials and Linear Control Systems.* New York, NY: Marcel Dekker, 1983.

BOLTYANSKII, VLADIMIR G. *Mathematical Methods of Optimal Control.* New York, NY: Holt, Rinehart and Winston, 1971.

CHUI, C.K. AND CHEN, G. *Linear Systems and Optimal Control.* New York, NY: Springer-Verlag, 1989.

JACOBS, O.L.R. *Introduction to Control Theory.* New York, NY: Oxford University Press, 1974.

* KAMIEN, MORTON I. AND SCHWARTZ, NANCY L. *Dynamic Optimization: The Calculus of Variations and Optimal Control in Economics and Management.* New York, NY: Elsevier Science, 1981.

LAGUNOV, V.N. *Introduction to Differential Games and Control Theory.* Berlin: Heldermann Verlag, 1985.

* LASALLE, JOSEPH P. *The Stability and Control of Discrete Processes.* New York, NY: Springer-Verlag, 1986.

LEITMANN, GEORGE. *The Calculus of Variations and Optimal Control: An Introduction.* New York, NY: Plenum Press, 1981.

PONTRJAGIN, LEV S., *et al.* *The Mathematical Theory of Optimal Processes.* New York, NY: Interscience, 1962.

TRAPPL, ROBERT, ED. *Cybernetics: Theory and Applications.* New York, NY: Hemisphere, 1983.

TU, PIERRE N.V. *Introductory Optimization Dynamics: Optimal Control with Economics and Management Science Applications.* New York, NY: Springer-Verlag, 1984.

** WONHAM, W. MURRAY. *Linear Multivariable Control: A Geometric Approach,* Third Edition. New York, NY: Springer-Verlag, 1985.

YOUNG, L.C. *Lectures on the Calculus of Variations and Optimal Control Theory.* New York, NY: W.B. Saunders, 1969; New York, NY: Chelsea, 1980.

20 Probability

20.1 General

BOREL, ÉMILE. *Elements of the Theory of Probability.* Englewood Cliffs, NJ: Prentice Hall, 1965.

* CACOULLOS, T. *Exercises in Probability.* New York, NY: Springer-Verlag, 1989.

* DAVID, F.N. *Games, Gods, and Gambling.* New York, NY: Hafner Press, 1962.

* GANI, J., ED. *The Craft of Probabilistic Modelling: A Collection of Personal Accounts.* New York, NY: Springer-Verlag, 1986.

** GNANADESIKAN, MRUDULLA; SCHEAFFER, RICHARD L.; AND SWIFT, JIM. *The Art and Techniques of Simulation.* Palo Alto, CA: Dale Seymour, 1987.

HUFF, DARRELL AND GEIS, I. *How to Take a Chance.* New York, NY: W.W. Norton, 1959.

* KAHNEMAN, DANIEL; SLOVIC, PAUL; AND TVERSKY, AMOS, EDS. *Judgment Under Uncertainty: Heuristics and Biases.* New York, NY: Cambridge University Press, 1982.

** MOSTELLER, FREDERICK. *Fifty Challenging Problems in Probability with Solutions.* Mineola, NY: Dover, 1987.

** NEWMAN, CLAIRE M.; OBREMSKI, THOMAS E.; AND SCHEAFFER, RICHARD L. *Exploring Probability.* Palo Alto, CA: Dale Seymour, 1987.

*** PACKEL, EDWARD W. *The Mathematics of Games and Gambling.* Washington, DC: Mathematical Association of America, 1981.

RÉNYI, ALFRÉD. *Letters on Probability.* Detroit, MI: Wayne State University Press, 1973.

THORP, EDWARD O. *Beat the Dealer: A Winning Strategy for the Game of Twenty-One.* New York, NY: Vintage Books, 1966.

WAGENAAR, WILLEM A. *Paradoxes of Gambling Behaviour.* Hillsdale, NJ: Lawrence Erlbaum, 1988.

*** WEAVER, WARREN. *Lady Luck: The Theory of Probability.* Mineola, NY: Dover, 1982.

20.2 Elementary

BRÉMAUD, PIERRE. *An Introduction to Probabilistic Modeling.* New York, NY: Springer-Verlag, 1988.

*** CHUNG, KAI LAI. *Elementary Probability Theory with Stochastic Processes,* Third Edition. New York, NY: Springer-Verlag, 1974, 1979.

CRAMÉR, HARALD. *The Elements of Probability Theory,* Second Edition. Melbourne, FL: Robert E. Krieger, 1973.

* GNEDENKO, BORIS V. AND KHINCHIN, A. YA. *An Elementary Introduction to the Theory of Probability.* Mineola, NY: Dover, 1962.

HODGES, JOSEPH L. AND LEHMANN, E.L. *Elements of Finite Probability,* Second Edition. San Francisco, CA: Holden-Day, 1970.

HOEL, PAUL G.; PORT, SIDNEY C.; AND STONE, CHARLES J. *Introduction to Probability Theory.* Boston, MA: Houghton Mifflin, 1971.

* KHAZANIE, RAMAKANT. *Basic Probability Theory and Applications.* Santa Monica, CA: Goodyear, 1976.

MOSTELLER, FREDERICK, *et al.* *Probability with Statistical Applications,* Second Edition. Reading, MA: Addison-Wesley, 1970.

OLKIN, INGRAM; GLESER, L.J.; AND DERMAN, C. *Probability Models and Applications.* New York, NY: Macmillan, 1978.

ROSS, SHELDON M. *Introduction to Probability Models,* Third Edition. New York, NY: Academic Press, 1972, 1985.

** SCHEAFFER, RICHARD L. *Introduction to Probability and Its Applications.* Boston, MA: PWS-Kent, 1990.

** SNELL, J. LAURIE. *Introduction to Probability.* Cambridge, MA: Random House, 1988.

* TUCKWELL, HENRY C. *Elementary Applications of Probability Theory.* New York, NY: Chapman and Hall, 1988.

20.3 Advanced

BREIMAN, LEO. *Probability*. Reading, MA: Addison-Wesley, 1968.

* CHOW, YUAN SHIH AND TEICHER, HENRY. *Probability Theory: Independence, Interchangeability, Martingales*, Second Edition. New York, NY: Springer-Verlag, 1988.

** CHUNG, KAI LAI. *A Course in Probability Theory*, Second Edition. New York, NY: Harcourt, Brace and World, 1968; New York, NY: Academic Press, 1974.

CRAMÉR, HARALD. *Random Variables and Probability Distributions*, Third Edition. New York, NY: Cambridge University Press, 1970.

DURRETT, RICHARD. *Probability: Theory and Examples*. Belmont, CA: Wadsworth, 1991.

*** FELLER, WILLIAM. *An Introduction to Probability Theory and Its Applications*, 2 Vols., Second Edition. New York, NY: John Wiley, 1968, 1971.

* GNEDENKO, BORIS V. *The Theory of Probability and the Elements of Statistics*, Fifth Edition. Moscow: MIR, 1976; New York, NY: Chelsea, 1967, 1989.

HUNTER, JEFFREY J. *Mathematical Techniques of Applied Probability*. New York, NY: Academic Press, 1983.

JOHNSON, NORMAN L. AND KOTZ, SAMUEL. *Urn Models and Their Application: An Approach to Modern Discrete Probability Theory*. New York, NY: John Wiley, 1977.

** KAC, MARK. *Statistical Independence in Probability, Analysis, and Number Theory*. Washington, DC: Mathematical Association of America, 1959.

* LLOYD, EMLYN. *Probability*. Handbook of Applicable Mathematics, Volume II. New York, NY: John Wiley, 1980.

** LOÈVE, MICHEL M. *Probability Theory*, 2 Vols., Fourth Edition. New York, NY: Van Nostrand Reinhold, 1963; New York, NY: Springer-Verlag, 1978.

MORAN, PATRICK A. *An Introduction to Probability Theory*. New York, NY: Oxford University Press, 1968, 1984.

PARZEN, EMANUEL. *Modern Probability Theory and Its Applications*. New York, NY: John Wiley, 1960.

RAO, M.M. *Probability Theory with Applications*. New York, NY: Academic Press, 1984.

* ROSENBLATT, MURRAY, ED. *Studies in Probability Theory*. Washington, DC: Mathematical Association of America, 1978.

*** ROTHSCHILD, V. AND LOGOTHETIS, N. *Probability Distributions*. New York, NY: John Wiley, 1986.

SHIRYAYEV, A.N. *Probability*. New York, NY: Springer-Verlag, 1984.

SPENCER, JOEL H. *Ten Lectures on the Probabilistic Method*. Philadelphia, PA: Society for Industrial and Applied Mathematics, 1987.

TRIVEDI, KISHOR S. *Probability and Statistics with Reliability, Queueing, and Computer Science Applications*. Englewood Cliffs, NJ: Prentice Hall, 1982.

* TUCKER, HOWARD G. *A Graduate Course in Probability*. New York, NY: Academic Press, 1967.

20.4 Stochastic Processes

BAILEY, NORMAN T.J. *The Elements of Stochastic Processes with Applications to the Natural Sciences*. New York, NY: John Wiley, 1990.

* BHAT, U. NARAYAN. *Elements of Applied Stochastic Processes*, Second Edition. New York, NY: John Wiley, 1984.

BROCKWELL, PETER J. AND DAVIS, RICHARD A. *Time Series: Theory and Methods*, Second Edition. New York, NY: Springer-Verlag, 1987, 1991.

CHOW, YUAN SHIH; ROBBINS, HERBERT; AND SIEGMUND, DAVID. *Great Expectations: The Theory of Optimal Stopping*. Boston, MA: Houghton Mifflin, 1971; Mineola, NY: Dover, 1991.

CHUNG, KAI LAI AND WILLIAMS, R.J. *Introduction to Stochastic Integration*, Second Edition. New York, NY: Birkhäuser, 1983, 1990.

COX, D.R. AND ISHAM, VALERIE. *Point Processes*. New York, NY: Chapman and Hall, 1980.

COX, D.R. AND SMITH, W.L. *Queues*. New York, NY: Halsted Press, 1961.

CRAMÉR, HARALD AND LEADBETTER, M.R. *Stationary and Related Stochastic Processes.* New York, NY: John Wiley, 1970.

* DELLACHERIE, CLAUDE AND MEYER, PAUL-ANDRÉ. *Probabilities and Potential.* Amsterdam: North-Holland, 1978.

*** DOOB, J.L. *Stochastic Processes.* New York, NY: John Wiley, 1965, 1990.

* DOYLE, PETER G. AND SNELL, J. LAURIE. *Random Walks and Electric Networks.* Washington, DC: Mathematical Association of America, 1984.

DUBINS, LESTER E. AND SAVAGE, LEONARD J. *Inequalities for Stochastic Processes (How to Gamble If You Must).* New York, NY: McGraw-Hill, 1965; Mineola, NY: Dover, 1976.

* HIDA, T. *Brownian Motion.* New York, NY: Springer-Verlag, 1980.

** KALLINPUR, GOPINATH. *Stochastic Filtering Theory.* New York, NY: Springer-Verlag, 1980.

* KARLIN, SAMUEL AND TAYLOR, HOWARD M. *A First Course in Stochastic Processes,* Second Edition. New York, NY: Academic Press, 1975.

KARLIN, SAMUEL AND TAYLOR, HOWARD M. *A Second Course in Stochastic Processes.* New York, NY: Academic Press, 1981.

KEMENY, JOHN G. AND SNELL, J. LAURIE. *Finite Markov Chains.* New York, NY: Springer-Verlag, 1976.

KEMENY, JOHN G.; SNELL, J. LAURIE; AND KNAPP, ANTHONY W. *Denumerable Markov Chains,* Second Edition. New York, NY: Springer-Verlag, 1976.

LUKACS, EUGENE. *Stochastic Convergence,* Second Edition. New York, NY: Academic Press, 1975.

* MEDHI, J. *Stochastic Processes.* New York, NY: Halsted Press, 1982.

* NELSON, EDWARD. *Radically Elementary Probability Theory.* Princeton, NJ: Princeton University Press, 1987.

PORT, SIDNEY C. AND STONE, CHARLES J. *Brownian Motion and Classical Potential Theory.* New York, NY: Academic Press, 1978.

ROSENBLATT, MURRAY. *Random Processes,* Second Edition. New York, NY: Springer-Verlag, 1974.

** ROSS, SHELDON M. *Stochastic Processes.* New York, NY: John Wiley, 1983.

SKOROKHOD, A.V. *Studies in the Theory of Random Processes.* Mineola, NY: Dover, 1982.

* SOBOL, I.M. *The Monte Carlo Method.* Chicago, IL: University of Chicago Press, 1974; Moscow: MIR, 1975.

SPITZER, FRANK. *Principles of Random Walk.* New York, NY: Van Nostrand Reinhold, 1964.

SYSKI, R. *Random Processes: A First Look,* Second Edition. New York, NY: Marcel Dekker, 1979, 1989.

* TAYLOR, HOWARD M. AND KARLIN, SAMUEL. *An Introduction to Stochastic Modeling.* New York, NY: Academic Press, 1984.

TOPSØE, FLEMMING. *Spontaneous Phenomena: A Mathematical Analysis.* New York, NY: Academic Press, 1990.

20.5 Foundations

* DE FINETTI, BRUNO. *Probability, Induction, and Statistics.* New York, NY: John Wiley, 1972.

** DE FINETTI, BRUNO. *Theory of Probability: A Critical Introductory Treatment,* 2 Vols. New York, NY: John Wiley, 1974, 1975.

FINE, TERRENCE L. *Theories of Probability: An Examination of Foundations.* New York, NY: Academic Press, 1973.

GOOD, I.J. *Good Thinking: The Foundations of Probability and Its Applications.* Minneapolis, MN: University of Minnesota Press, 1983.

HARPER, W.L. AND HOOKER, C.A., EDS. *Foundations of Probability Theory, Statistical Inference, and Statistical Theories of Sciences,* 2 Vols. Norwell, MA: D. Reidel, 1976.

** KOLMOGOROV, ANDREI N. *Foundations of the Theory of Probability,* Second Edition. New York, NY: Chelsea, 1950, 1956.

RÉNYI, ALFRÉD. *Foundations of Probability.* San Francisco, CA: Holden-Day, 1970.

* SHAFER, GLENN. *A Mathematical Theory of Evidence.* Princeton, NJ: Princeton University Press, 1976.

*** VON MISES, RICHARD. *Probability, Statistics, and Truth,* Second Revised English Edition. New York, NY: Macmillan, 1957; Mineola, NY: Dover, 1981.

20.6 Measure Theory

ASH, ROBERT B. *Real Analysis and Probability.* New York, NY: Academic Press, 1972.

BAUER, HEINZ. *Probability Theory and Elements of Measure Theory,* Second English Edition. New York, NY: Academic Press, 1978, 1981.

* BILLINGSLEY, PATRICK. *Convergence of Probability Measures.* New York, NY: John Wiley, 1968.

** BILLINGSLEY, PATRICK. *Probability and Measure,* Second Edition. New York, NY: John Wiley, 1979, 1986.

DUDLEY, RICHARD M. *Real Analysis and Probability.* Belmont, CA: Wadsworth, 1989.

KALLENBERG, OLAV. *Random Measures.* New York, NY: Academic Press, 1983.

* KINGMAN, J.F.C. AND TAYLOR, S.J. *Introduction to Measure with Probability.* New York, NY: Cambridge University Press, 1966.

* KRIEGER, HENRY A. *Measure-theoretic Probability.* Lanham, MD: University Press of America, 1980.

** PARTHASARATHY, K.R. *Probability Measures on Metric Spaces.* New York, NY: Academic Press, 1967.

20.7 Fuzzy Sets

KAUFMANN, ARNOLD AND GUPTA, MADAN M. *Introduction to Fuzzy Arithmetic: Theory and Applications.* New York, NY: Van Nostrand Reinhold, 1985.

* KAUFMANN, ARNOLD. *Introduction to the Theory of Fuzzy Subsets,* New York, NY: Academic Press, 1975.

KLIR, GEORGE J. AND FOLGER, TINA A. *Fuzzy Sets, Uncertainty, and Information.* Englewood Cliffs, NJ: Prentice Hall, 1988.

YAGER, R.R. *Fuzzy Sets and Possibility Theory.* Elmsford, NY: Pergamon Press, 1982.

20.8 Special Topics

BHARUCHA-REID, A.T. *Probabilistic Methods in Applied Mathematics,* 2 Vols. New York, NY: Academic Press, 1968, 1970.

DOOB, J.L. *Classical Potential Theory and Its Probabilistic Counterpart.* New York, NY: Springer-Verlag, 1984.

** GRAY, ROBERT M. *Probability, Random Processes, and Ergodic Properties.* New York, NY: Springer-Verlag, 1988.

GRENANDER, ULF. *Probabilities on Algebraic Structures.* New York, NY: John Wiley, 1963.

* GUDDER, STANLEY P. *Quantum Probability.* New York, NY: Academic Press, 1988.

** ROMANO, JOSEPH P. AND SIEGEL, ANDREW F. *Counterexamples in Probability and Statistics.* Belmont, CA: Wadsworth, 1986.

* SOLOMON, HERBERT. *Geometric Probability.* Philadelphia, PA: Society for Industrial and Applied Mathematics, 1978.

SPRINGER, M.D. *The Algebra of Random Variables.* New York, NY: John Wiley, 1979.

** SZÉKELY, GÁBOR J. *Paradoxes in Probability Theory and Mathematical Statistics.* Norwell, MA: D. Reidel, 1986.

* YAGLOM, A.M. AND YAGLOM, I.M. *Probability and Information.* Norwell, MA: D. Reidel, 1983.

21 Statistics

21.1 General

* BROOK, RICHARD J., *et al.*, EDS. *The Fascination of Statistics.* New York, NY: Marcel Dekker, 1986.
CAMPBELL, STEPHEN K. *Flaws and Fallacies in Statistical Thinking.* Englewood Cliffs, NJ: Prentice Hall, 1973.
* DEGROOT, MORRIS H.; FIENBERG, STEPHEN E.; AND KADANE, JOSEPH B., EDS. *Statistics and the Law.* New York, NY: John Wiley, 1986.
** DEMING, W. EDWARDS. *Out of The Crisis.* Cambridge, MA: MIT Press, 1986.
* FINKELSTEIN, MICHAEL O. AND LEVIN, BRUCE. *Statistics for Lawyers.* New York, NY: Springer-Verlag, 1990.
FOLKS, J. LEROY. *Ideas of Statistics.* New York, NY: John Wiley, 1981.
HACKING, IAN. *Logic of Statistical Inference.* New York, NY: Cambridge University Press, 1976.
HAND, D.J. AND EVERITT, B.S., EDS. *The Statistical Consultant in Action.* New York, NY: Cambridge University Press, 1987.
HOGG, ROBERT V., ED. *Studies in Statistics.* Washington, DC: Mathematical Association of America, 1978.
HOLLANDER, MYLES AND PROSCHAN, FRANK. *The Statistical Exorcist: Dispelling Statistics Anxiety.* New York, NY: Marcel Dekker, 1984.
HOOKE, ROBERT. *How to Tell the Liars from the Statisticians.* New York, NY: Marcel Dekker, 1983.
HUFF, DARRELL AND GEIS, I. *How to Lie with Statistics.* New York, NY: W.W. Norton, 1954.
* JAFFE, A.J. AND SPIRER, HERBERT F. *Misused Statistics: Straight Talk for Twisted Numbers.* New York, NY: Marcel Dekker, 1987.
** LANDWEHR, JAMES M. AND WATKINS, ANN E. *Exploring Data.* Palo Alto, CA: Dale Seymour, 1986.
** LANDWEHR, JAMES M.; SWIFT, JIM; AND WATKINS, ANN E. *Exploring Surveys and Information From Samples.* Palo Alto, CA: Dale Seymour, 1987.
MORONEY, M.J. *Facts From Figures.* New York, NY: Penguin, 1957.
PHILLIPS, JOHN L., JR. *How to Think about Statistics.* New York, NY: W.H. Freeman, 1988.
*** TANUR, JUDITH M. AND MOSTELLER, FREDERICK, EDS. *Statistics: A Guide to the Unknown,* Third Edition. San Francisco, CA: Holden-Day, 1972; Belmont, CA: Wadsworth, 1989.
** TUFTE, EDWARD R. *Envisioning Information.* Cheshire, CT: Graphics Press, 1990.
*** TUFTE, EDWARD R. *The Visual Display of Quantitative Information.* Cheshire, CT: Graphics Press, 1983.
WADSWORTH, HARRISON M., ED. *Handbook of Statistical Methods for Engineers and Scientists.* New York, NY: McGraw-Hill, 1990.

21.2 Introductory Texts

ARNEY, WILLIAM R. *Understanding Statistics in the Social Sciences.* New York, NY: W.H. Freeman, 1990.
* DEVORE, JAY L. AND PECK, ROXY. *Statistics: The Exploration and Analysis of Data.* St. Paul, MN: West, 1986.
*** FREEDMAN, DAVID, *et al.* *Statistics,* Second Edition. New York, NY: W.W. Norton, 1978, 1991.
GROENVELD, RICHARD A. *Introductory Statistical Methods.* Boston, MA: PWS-Kent, 1988.
KACHIGAN, SAM K. *Statistical Analysis: An Interdisciplinary Introduction to Univariate and Multivariate Methods.* New York, NY: Radius Press, 1986.
** KOOPMANS, LAMBERT H. *Introduction to Contemporary Statistical Methods,* Second Edition. Boston, MA: PWS-Kent, 1987.
MALLIOS, WILLIAM S. *Statistical Modeling: Applications in Contemporary Issues.* Ames, IA: Iowa State University Press, 1989.
* MCCLAVE, JAMES T. AND DIETRICH, FRANK H. *Statistics,* Fifth Edition. San Francisco, CA: Dellen, 1982, 1991.

** MOORE, DAVID S. AND MCCABE, GEORGE P. *Introduction to the Practice of Statistics.* New York, NY: W.H. Freeman, 1989.

*** MOORE, DAVID S. *Statistics: Concepts and Controversies,* Third Edition. New York, NY: W.H. Freeman, 1979, 1991.

NOETHER, GOTTFRIED. *Introduction to Statistics: A Fresh Approach.* Boston, MA: Houghton Mifflin, 1971; New York, NY: Springer-Verlag, 1991.

* PETERS, WILLIAM S. *Counting for Something: Statistical Principles and Personalities.* New York, NY: Springer-Verlag, 1986.

** SIEGEL, ANDREW F. *Statistics and Data Analysis: An Introduction.* New York, NY: John Wiley, 1988.

21.3 Elementary

BLALOCK, HUBERT M., JR. *Social Statistics,* Second Edition. New York, NY: McGraw-Hill, 1972.

* BROWN, BYRON WM., JR. AND HOLLANDER, MYLES. *Statistics: A Biomedical Introduction.* New York, NY: John Wiley, 1977.

CHATFIELD, CHRISTOPHER. *Problem Solving: A Statistician's Guide.* New York, NY: Chapman and Hall, 1988.

COX, C. PHILIP. *A Handbook of Introductory Statistical Methods.* New York, NY: John Wiley, 1987.

HUCK, SCHUYLER W. AND SANDLER, HOWARD M. *Statistical Illusions: Problems.* New York, NY: Harper and Row, 1984.

* LARSEN, RICHARD J. AND STROUP, DONNA F. *Statistics in the Real World: A Book of Examples.* New York, NY: Macmillan, 1976.

* MOSES, LINCOLN E. *Think and Explain with Statistics.* Reading, MA: Addison-Wesley, 1986.

MOSTELLER, FREDERICK AND ROURKE, ROBERT E.K. *Sturdy Statistics: Nonparametrics and Order Statistics.* Reading, MA: Addison-Wesley, 1973.

* YOUDEN, W.J. *Risk, Choice, and Prediction: An Introduction to Experimentation.* Boston, MA: Duxbury Press, 1974.

YOUDEN, W.J. *Experimentation and Measurement.* Washington, DC: National Bureau of Standards, 1984.

21.4 Intermediate

* BICKEL, PETER J. AND DOKSUM, KJELL A. *Mathematical Statistics: Basic Ideas and Selected Topics.* Oakland, CA: Holden-Day, 1977.

*** BOX, GEORGE E.P.; HUNTER, WILLIAM G.; AND HUNTER, J. STUART. *Statistics for Experimenters: An Introduction to Design, Data Analysis, and Model Building.* New York, NY: John Wiley, 1978.

* CASELLA, GEORGE AND BERGER, ROGER L. *Statistical Inference.* Belmont, CA: Wadsworth, 1990.

CHERNOFF, HERMAN AND MOSES, LINCOLN E. *Elementary Decision Theory.* New York, NY: John Wiley, 1959; Mineola, NY: Dover, 1986.

COX, D.R. AND SNELL, E. JOYCE. *Applied Statistics: Principles and Examples.* New York, NY: Chapman and Hall, 1981.

* DEGROOT, MORRIS H. *Probability and Statistics,* Second Edition. Reading, MA: Addison-Wesley, 1975, 1986.

DEVORE, JAY L. *Probability and Statistics for Engineering and the Sciences,* Third Edition. Pacific Grove, CA: Brooks/Cole, 1982, 1991.

HOGG, ROBERT V. AND CRAIG, ALLEN T. *Introduction to Mathematical Statistics,* Fourth Edition. New York, NY: Macmillan, 1958, 1978.

HOGG, ROBERT V. AND LEDOLTER, JOHANNES. *Engineering Statistics.* New York, NY: Macmillan, 1987.

** HOGG, ROBERT V. AND TANIS, ELLIOT A. *Probability and Statistical Inference,* Third Edition. New York, NY: Macmillan, 1977, 1988.

KALBFLEISCH, J.G. *Probability and Statistical Inference,* 2 Vols., Second Edition. New York, NY: Springer-Verlag, 1979, 1985.

KEMPTHORNE, OSCAR AND FOLKS, J. LEROY. *Probability, Statistics, and Data Analysis.* Ames, IA: Iowa State University Press, 1971.

* KIEFER, JACK C. *Introduction to Statistical Inference.* New York, NY: Springer-Verlag, 1987.

* LARSEN, RICHARD J. AND MARX, MORRIS L. *An Introduction to Mathematical Statistics and Its Applications,* Second Edition. Englewood Cliffs, NJ: Prentice Hall, 1981, 1986.

* MADANSKY, ALBERT. *Prescriptions for Working Statisticians.* New York, NY: Springer-Verlag, 1988.

MANOUKIAN, EDWARD B. *Modern Concepts and Theorems of Mathematical Statistics.* New York, NY: Springer-Verlag, 1986.

MENDENHALL, WILLIAM; WACKERLY, DENNIS D.; AND SCHEAFFER, RICHARD L. *Mathematical Statistics with Applications,* Fourth Edition. Boston, MA: PWS-Kent, 1990.

MOOD, ALEXANDER M.; GRAYBILL, FRANKLIN A.; AND BOES, D.C. *Introduction to the Theory of Statistics,* Third Edition. New York, NY: McGraw-Hill, 1974.

** RAO, C. RADHKIRSHNA. *Linear Statistical Inference and Its Applications,* Second Edition. New York, NY: John Wiley, 1973.

* RICE, JOHN A. *Mathematical Statistics and Data Analysis.* Belmont, CA: Wadsworth, 1988.

SCHEAFFER, RICHARD L. AND MCCLAVE, JAMES T. *Probability and Statistics for Engineers,* Third Edition. Boston, MA: Duxbury Press, 1990.

*** SNEDECOR, GEORGE W. AND COCHRAN, WILLIAM G. *Statistical Methods,* Seventh Edition. Ames, IA: Iowa State University Press, 1967, 1980.

WETHERILL, G. BARRIE. *Intermediate Statistical Methods.* New York, NY: Chapman and Hall, 1981.

21.5 Advanced

COX, D.R. AND HINKLEY, DAVID V. *Theoretical Statistics.* New York, NY: Chapman and Hall, 1974.

* COX, D.R. AND OAKES, D. *Analysis of Survival Data.* New York, NY: Chapman and Hall, 1984.

DAVID, H.A. *Order Statistics,* Second Edition. New York, NY: John Wiley, 1981.

* EDGINGTON, EUGENE S. *Randomization Tests,* Second Edition. New York, NY: Marcel Dekker, 1980, 1987.

FARNUM, NICHOLAS R. AND STANTON, LAVERNE W. *Quantitative Forecasting Methods.* Boston, MA: PWS-Kent, 1989.

HAMPEL, FRANK R., *et al. Robust Statistics: The Approach Based on Influence Functions.* New York, NY: John Wiley, 1986.

** HINKLEY, DAVID V.; REID, NANCY; AND SNELL, E. JOYCE, EDS. *Statistical Theory and Modelling.* New York, NY: Chapman and Hall, 1990.

** JOHNSON, NORMAN L. AND KOTZ, SAMUEL. *Continuous Univariate Distributions,* 2 Vols. Boston, MA: Houghton Mifflin, 1970.

** JOHNSON, NORMAN L. AND KOTZ, SAMUEL. *Discrete Distributions.* Boston, MA: Houghton Mifflin, 1969.

*** KENDALL, MAURICE; STUART, ALAN; AND ORD, J. KEITH. *The Advanced Theory of Statistics,* 3 Vols. New York, NY: Hafner Press, 1973; New York, NY: Macmillan, 1979; New York, NY: Oxford University Press, 1987.

LEHMANN, E.L. *Theory of Point Estimation.* New York, NY: John Wiley, 1983; Belmont, CA: Wadsworth, 1991.

* LLOYD, EMLYN, ED. *Statistics.* Handbook of Applicable Mathematics, Volume VI. New York, NY: John Wiley, 1984.

21.6 Data Analysis

** AGRESTI, ALAN. *Categorical Data Analysis.* New York, NY: John Wiley, 1990.

* ANDREWS, D.F. AND HERZBERG, A.M. *Data: A Collection of Problems from Many Fields for the Student and Research Worker.* New York, NY: Springer-Verlag, 1985.

BREIMAN, LEO, *et al. Classification and Regression Trees.* Belmont, CA: Wadsworth, 1984.

CHAMBERS, JOHN M., *et al. Graphical Methods for Data Analysis.* Belmont, CA: Wadsworth, 1983.

*** CLEVELAND, WILLIAM S. *The Elements of Graphing Data,* Second Edition. Belmont, CA: Wadsworth, 1985, 1991.

** COX, D.R. AND SNELL, E. JOYCE. *Analysis of Binary Data,* Second Edition. New York, NY: Chapman and Hall, 1989.

* EFRON, BRADLEY. *The Jackknife, the Bootstrap, and Other Resampling Plans.* Philadelphia, PA: Society for Industrial and Applied Mathematics, 1982.

ERICKSON, BONNIE H. AND NOSANCHUK, T.A. *Understanding Data.* New York, NY: McGraw-Hill, 1977.

* FIENBERG, STEPHEN E. *The Analysis of Cross-Classified Categorical Data.* Cambridge, MA: MIT Press, 1977.

FLEISS, JOSEPH L. *Statistical Methods for Rates and Proportions,* Second Edition. New York, NY: John Wiley, 1981.

* HOAGLIN, DAVID C.; MOSTELLER, FREDERICK; AND TUKEY, JOHN W., EDS. *Exploring Data Tables, Trends, and Shapes.* New York, NY: John Wiley, 1985.

*** HOAGLIN, DAVID C.; MOSTELLER, FREDERICK; AND TUKEY, JOHN W., EDS. *Understanding Robust and Exploratory Data Analysis.* New York, NY: John Wiley, 1983.

* MOSTELLER, FREDERICK; FIENBERG, STEPHEN E.; AND ROURKE, ROBERT E.K. *Beginning Statistics with Data Analysis.* Reading, MA: Addison-Wesley, 1983.

TUKEY, JOHN W. *Exploratory Data Analysis.* Reading, MA: Addison-Wesley, 1977.

VELLEMAN, PAUL F. AND HOAGLIN, DAVID C. *Applications, Basics, and Computing of Exploratory Data Analysis.* Boston, MA: Duxbury Press, 1981.

21.7 Bayesian Methods

* BERGER, JAMES O. *Statistical Decision Theory and Bayesian Analysis,* Second Edition. New York, NY: Springer-Verlag, 1980, 1985.

BOX, GEORGE E.P. AND TIAO, GEORGE C. *Bayesian Inference in Statistical Analysis.* Reading, MA: Addison-Wesley, 1973.

HOWSON, COLIN AND URBACH, PETER. *Scientific Reasoning: The Bayesian Approach.* La Salle, IL: Open Court, 1989.

IVERSON, GUDMUND R. *Bayesian Statistical Inference.* Beverly Hills, CA: Sage, 1984.

LINDLEY, D.V. *Bayesian Statistics: A Review.* Philadelphia, PA: Society for Industrial and Applied Mathematics, 1971.

* PRESS, S. JAMES. *Bayesian Statistics: Principles, Models, and Applications.* New York, NY: John Wiley, 1989.

21.8 Computational Statistics

BECKER, RICHARD A.; CHAMBERS, JOHN M.; AND WILKS, ALLAN R. *The New S Language: A Programming Environment for Data Analysis and Graphics.* Belmont, CA: Wadsworth, 1988.

HAMMERSLEY, J.M. AND HANDSCOMB, D.C. *Monte Carlo Methods.* New York, NY: John Wiley, 1964.

KENNEDY, WILLIAM J., JR. AND GENTLE, JAMES E. *Statistical Computing.* New York, NY: Marcel Dekker, 1980.

* THISTED, RONALD A. *Elements of Statistical Computing: Numerical Computation.* New York, NY: Chapman and Hall, 1988.

21.9 Linear Models and Regression Analysis

ATKINSON, ANTHONY C. *Plots, Transformations, and Regression: An Introduction to Graphical Methods of Diagnostic Regression Analysis.* New York, NY: Clarendon Press, 1985.

* BATES, DOUGLAS M. AND WATTS, DONALD G. *Nonlinear Regression Analysis and Its Applications.* New York, NY: John Wiley, 1988.

BELSLEY, DAVID A.; KUH, EDWIN; AND WELSCH, ROY E. *Regression Diagnostics: Identifying Influential Data and Sources of Collinearity.* New York, NY: John Wiley, 1980.

CHATTERJEE, SAMPRIT AND PRICE, BERTRAM. *Regression Analysis by Example.* New York, NY: John Wiley, 1977.

* COOK, R. DENNIS AND WEISBERG, SANFORD. *Residuals and Influence in Regression.* New York, NY: Chapman and Hall, 1982.

DOBSON, ANNETTE J. *An Introduction to Generalized Linear Models.* New York, NY: Chapman and Hall, 1990.

** DRAPER, NORMAN R. AND SMITH, H. *Applied Regression Analysis,* Second Edition. New York, NY: John Wiley, 1981.

GRAYBILL, FRANKLIN A. *Theory and Application of the Linear Model.* Boston, MA: Duxbury Press, 1976.

GUNST, RICHARD F. AND MASON, ROBERT L. *Regression Analysis and Its Application: A Data-Oriented Approach.* New York, NY: Marcel Dekker, 1980.

HASTIE, T. AND TIBSHIRANI, R.J. *Generalized Additive Models.* New York, NY: Chapman and Hall, 1990.

* MCCULLAGH, P. AND NELDER, J.A. *Generalized Linear Models,* Second Edition. New York, NY: Chapman and Hall, 1983, 1989.

MORRISON, DONALD F. *Applied Linear Statistical Methods.* Englewood Cliffs, NJ: Prentice Hall, 1983.

** MOSTELLER, FREDERICK AND TUKEY, JOHN W. *Data Analysis and Regression: A Second Course in Statistics.* Reading, MA: Addison-Wesley, 1977.

* NETER, JOHN; WASSERMAN, WILLIAM; AND KUTNER, MICHAEL H. *Applied Linear Statistical Models: Regression, Analysis of Variance, and Experimental Designs,* Second Edition. Homewood, IL: Richard D. Irwin, 1983, 1985.

*** WEISBERG, SANFORD. *Applied Linear Regression,* Second Edition. New York, NY: John Wiley, 1980, 1985.

21.10 Multivariate Analysis

ANDERSON, T.W. *An Introduction to Multivariate Statistical Analysis,* Second Edition. New York, NY: John Wiley, 1984.

** BISHOP, YVONNE M.M., *et al. Discrete Multivariate Analysis: Theory and Practice.* Cambridge, MA: MIT Press, 1977.

CHATFIELD, CHRISTOPHER AND COLLINS, ALEXANDER J. *Introduction to Multivariate Analysis.* New York, NY: Chapman and Hall, 1980.

* GNANADESIKAN, R. *Methods For Statistical Data Analysis of Multivariate Observations.* New York, NY: John Wiley, 1977.

GORDON, A.D. *Classification: Methods for the Exploratory Analysis of Multivariate Data.* New York, NY: Chapman and Hall, 1981.

GREEN, PAUL E. AND CARROLL, J. DOUGLAS. *Mathematical Tools for Applied Multivariate Analysis.* New York, NY: Academic Press, 1976.

HAND, D.J. *Discrimination and Classification.* New York, NY: John Wiley, 1981.

JAMBU, MICHEL. *Exploratory and Multivariate Data Analysis.* New York, NY: Academic Press, 1991.

JOBSON, J.D. *Applied Multivariate Data Analysis.* New York, NY: Springer-Verlag, 1991.

** JOHNSON, RICHARD A. AND WICHERN, DEAN W. *Applied Multivariate Statistical Analysis,* Second Edition. Englewood Cliffs, NJ: Prentice Hall, 1982, 1988.

KENDALL, MAURICE. *Multivariate Analysis,* Second Edition. New York, NY: Macmillan, 1980.

* MORRISON, DONALD F. *Multivariate Statistical Methods,* Third Edition. New York, NY: McGraw-Hill, 1967, 1990.

21.11 Nonparametric Statistics

GIBBONS, JEAN D. *Nonparametric Methods for Quantitative Analysis*, Second Edition. Syracuse, NY: American Sciences Press, 1985.

** LEHMANN, E.L. AND D'ABRERA, H.J.M. *Nonparametrics: Statistical Methods Based on Ranks.* San Francisco, CA: Holden-Day, 1975.

* RANDLES, RONALD H. AND WOLFE, DOUGLAS A. *Introduction to the Theory of Nonparametric Statistics.* New York, NY: John Wiley, 1979.

SPRENT, PETER. *Applied Nonparametric Statistical Methods.* New York, NY: Chapman and Hall, 1989.

21.12 Experimental Design

** FISHER, RONALD A. *Statistical Methods, Experimental Design, and Scientific Inference*, Combined Edition. New York, NY: Hafner Press, 1959; New York, NY: Oxford University Press, 1990.

* MASON, ROBERT L.; GUNST, RICHARD F.; AND HESS, JAMES L. *Statistical Design and Analysis of Experiments: With Applications to Engineering and Science.* New York, NY: John Wiley, 1989.

* MONTGOMERY, DOUGLAS C. *Design and Analysis of Experiments*, Third Edition. New York, NY: John Wiley, 1984, 1991.

21.13 Sampling and Survey Design

COCHRAN, WILLIAM G. *Sampling Techniques*, Third Edition. New York, NY: John Wiley, 1977.

* DEMING, W. EDWARDS. *Some Theory of Sampling.* Mineola, NY: Dover, 1966.

** SCHEAFFER, RICHARD L.; MENDENHALL, WILLIAM; AND OTT, LYMAN. *Elementary Survey Sampling.* Belmont, CA: Wadsworth, 1971; Boston, MA: PWS-Kent, 1990.

SUKHATME, P.V. AND SUKHATME, B.V. *Sampling Theory of Surveys with Applications*, Second Edition. Ames, IA: Iowa State University Press, 1970.

WAINER, HOWARD, ED. *Drawing Inferences from Self-Selected Samples.* New York, NY: Springer-Verlag, 1986.

* WILLIAMS, BILL. *A Sampler on Sampling.* New York, NY: John Wiley, 1978.

21.14 Quality Control

BURR, IRVING W. *Elementary Statistical Quality Control.* New York, NY: Marcel Dekker, 1979.

DEHNAD, KHOSROW, ED. *Quality Control, Robust Design, and the Taguchi Method.* Belmont, CA: Wadsworth, 1989.

* MONTGOMERY, DOUGLAS C. *Introduction to Statistical Quality Control*, Second Edition. New York, NY: John Wiley, 1985, 1991.

* RYAN, THOMAS P. *Statistical Methods for Quality Improvement.* New York, NY: John Wiley, 1989.

* WADSWORTH, HARRISON M.; STEPHENS, KENNETH S.; AND GODFREY, A. BLANTON. *Modern Methods For Quality Control and Improvement.* New York, NY: John Wiley, 1986.

21.15 Time Series

** BOX, GEORGE E.P. AND JENKINS, GWILYM M. *Time Series Analysis, Forecasting, and Control*, Revised Edition. Oakland, CA: Holden-Day, 1976.

* CHATFIELD, CHRISTOPHER. *The Analysis of Time Series: An Introduction*, Fourth Edition. (Former title: *The Analysis of Time Series: Theory and Practice.*) New York, NY: Chapman and Hall, 1975, 1989.

COX, D.R. AND LEWIS, P.A.W. *The Statistical Analysis of Series of Events.* New York, NY: Methuen, 1966.

CRYER, JONATHAN D. *Time Series Analysis.* Boston, MA: Duxbury Press, 1986.

* DIGGLE, PETER J. *Time Series: A Biostatistical Approach.* New York, NY: Oxford University Press, 1990.

21.16 Special Topics

BRAINERD, BARRON. *Weighing Evidence in Language and Literature: A Statistical Approach.* Toronto: University of Toronto Press, 1974.

BUNCHER, C. RALPH AND TSAY, JIA-YEONG, EDS. *Statistics in the Pharmaceutical Industry.* New York, NY: Marcel Dekker, 1981.

CLEARY, JAMES P. AND LEVENBACH, HANS. *The Professional Forecaster: The Forecasting Process Through Data Analysis.* Belmont, CA: Lifetime Learning, 1982.

GIBBONS, JEAN D.; OLKIN, INGRAM; AND SOBEL, MILTON. *Selecting and Ordering Populations: A New Statistical Methodology.* New York, NY: John Wiley, 1977.

HEDGES, L. AND OLKIN, INGRAM. *Statistical Methods for Meta-Analysis.* New York, NY: Academic Press, 1985.

ISAAKS, EDWARD H. AND SRIVASTAVA, R. MOHAN. *An Introduction to Applied Geostatistics.* New York, NY: Oxford University Press, 1990.

MACIEJOWSKI, JAN M. *The Modelling of Systems with Small Observation Sets.* New York, NY: Springer-Verlag, 1978.

PANKRATZ, ALAN. *Forecasting With Univariate Box-Jenkins Models: Concepts and Cases.* New York, NY: John Wiley, 1983.

22 Computer Science

22.1 Computer Literacy

ARGANBRIGHT, DEAN E. *Mathematical Applications of Electronic Spreadsheets.* New York, NY: McGraw-Hill, 1985.

** BIERMAN, ALAN. *Great Ideas in Computer Science.* Cambridge, MA: MIT Press, 1990.

BRUCE, J.W.; GIBLIN, P.J.; AND RIPPON, P.J. *Microcomputers and Mathematics.* New York, NY: Cambridge University Press, 1990.

** DECKER, RICK AND HIRSHFIELD, STUART. *The Analytical Engine.* Belmont, CA: Wadsworth, 1990.

*** DEWDNEY, A.K. *The Armchair Universe: An Exploration of Computer Worlds.* New York, NY: W.H. Freeman, 1988.

** DEWDNEY, A.K. *The Turing Omnibus: 61 Excursions in Computer Science.* Rockville, MD: Computer Science Press, 1989.

GRAY, THEODORE W. AND GLYNN, JERRY. *Exploring Mathematics with Mathematica: Dialogs Concerning Computers and Mathematics.* Reading, MA: Addison-Wesley, 1991.

* HAREL, DAVID. *Algorithmics: The Spirit of Computing.* Reading, MA: Addison-Wesley, 1987.

KIDDER, TRACY. *The Soul of a New Machine.* Waltham, MA: Little, Brown, 1981.

* MCCORDUCK, PAMELA. *Machines Who Think: A Personal Inquiry into the History and Prospects of Artificial Intelligence.* New York, NY: W.H. Freeman, 1979.

MOREAU, R. *The Computer Comes of Age: The People, the Hardware, and the Software.* Cambridge, MA: MIT Press, 1984.

SAVAGE, JOHN E.; MAGIDSON, SUSAN; AND STEIN, ALEX M. *The Mystical Machine: Issues and Ideas in Computing.* Reading, MA: Addison-Wesley, 1986.

* SHORE, JOHN. *The Sachertorte Algorithm and Other Antidotes to Computer Anxiety.* New York, NY: Viking Press, 1985.

* VON NEUMANN, JOHN. *The Computer and the Brain,* New Haven, CT: Yale University Press, 1979.

** WEISS, ERIC A., ED. *A Computer Science Reader: Selections from Abacus.* New York, NY: Springer-Verlag, 1988.

22.2 Computers and Society

* BOLTER, J. DAVID. *Turing's Man: Western Culture in the Computer Age.* Chapel Hill, NC: University of North Carolina Press, 1984.

DEJOIE, ROY M., *et al. Ethical Issues in Information Systems.* Boston, MA: Boyd and Fraser, 1991.

DREYFUS, HUBERT L. *What Computers Can't Do.* New York, NY: Harper and Row, 1972.

* DUNLAP, CHARLES AND KLING, ROB, EDS. *Computerization and Controversy: Value Conflicts and Social Choices.* New York, NY: Academic Press, 1991.

** GRAUBARD, STEPHEN R., ED. *The Artificial Intelligence Debate: False Starts, Real Foundations.* Cambridge, MA: MIT Press, 1988.

* JENNINGS, KARLA. *The Devouring Fungus: Tales of the Computer Age.* New York, NY: W.W. Norton, 1990.

KEMENY, JOHN G. *Man and the Computer.* New York, NY: Charles Scribner's, 1972.

* ROSZAK, THEODORE. *The Cult of Information: The Folklore of Computers and the True Art of Thinking.* Cambridge, MA: Pantheon Books, 1986.

** STOLL, CLIFFORD. *The Cuckoo's Egg: Tracking a Spy Through the Maze of Computer Espionage.* New York, NY: Doubleday, 1989.

** WEIZENBAUM, JOSEPH. *Computer Power and Human Reason: From Judgment to Calculation.* New York, NY: W.H. Freeman, 1976.

WIENER, NORBERT. *The Human Use of Human Beings: Cybernetics and Society.* New York, NY: Doubleday, 1954.

22.3 Introductory

** ABELSON, HAROLD; SUSSMAN, GERALD J.; AND SUSSMAN, JULIE. *Structure and Interpretation of Computer Programs.* Cambridge, MA: MIT Press, 1985.

* BROOKSHEAR, J. GLENN. *Computer Science: An Overview,* Second Edition. Redwood City, CA: Benjamin Cummings, 1985, 1988.

CARBERRY, M. SANDRA; COHEN, A. TONI; AND KHALIL, HATEM M. *Principles of Computer Science: Concepts, Algorithms, Data Structures, and Applications.* Rockville, MD: Computer Science Press, 1986.

* McGETTRICK, ANDREW D. AND SMITH, PETER D. *Graded Problems in Computer Science.* Reading, MA: Addison-Wesley, 1983.

POLLACK, SEYMOUR V., ED. *Studies in Computer Science.* Washington, DC: Mathematical Association of America, 1982.

* TUCKER, ALLEN B., *et al. Fundamentals of Computing I: Logic, Problem-Solving, Programs, and Computers.* New York, NY: McGraw-Hill, 1991.

WULF, WILLIAM A., *et al. Fundamental Structures of Computer Science.* Reading, MA: Addison-Wesley, 1981.

22.4 Data Structures

*** AHO, ALFRED V.; HOPCROFT, JOHN E.; AND ULLMAN, JEFFREY D. *Data Structures and Algorithms.* Reading, MA: Addison-Wesley, 1983.

DALE, NELL AND LILLY, SUSAN C. *Pascal Plus Data Structures.* Lexington, MA: D.C. Heath, 1988.

** HOROWITZ, ELLIS AND SAHNI, SARTAJ. *Fundamentals of Data Structures in Pascal,* Third Edition. Rockville, MD: Computer Science Press, 1976, 1990.

* KORSH, JAMES F. AND GARRETT, LEONARD J. *Data Structures, Algorithms, and Program Style Using C.* Boston, MA: PWS-Kent, 1986, 1988.

* KRUSE, ROBERT L. *Data Structures and Program Design,* Second Edition. Englewood Cliffs, NJ: Prentice Hall, 1984, 1987.

* MEHLHORN, KURT. *Data Structures and Algorithms,* 3 Vols. New York, NY: Springer-Verlag, 1984.

MILLER, NANCY E. *File Structures Using Pascal.* Redwood City, CA: Benjamin Cummings, 1987.

SMITH, HARRY F. *Data Structures: Form and Function.* San Diego, CA: Harcourt Brace Jovanovich, 1987.

STUBBS, DAVID F. AND WEBRE, NEIL W. *Data Structures,* Second Edition. Pacific Grove, CA: Brooks/Cole, 1989.

WELSH, JIM; ELDER, JOHN; AND BUSTARD, DAVID. *Sequential Program Structures.* Englewood Cliffs, NJ: Prentice Hall, 1984.

** WIRTH, NIKLAUS. *Algorithms and Data Structures.* (Former title: *Algorithms + Data Structures = Programs.*) Englewood Cliffs, NJ: Prentice Hall, 1976, 1986.

22.5 Database Systems

ALAGIĆ, SUAD. *Relational Database Technology.* New York, NY: Springer-Verlag, 1986.

** CODD, E.F. *The Relational Model.* Reading, MA: Addison-Wesley, 1990.

*** DATE, C.J. *An Introduction to Database Systems,* Fourth Edition. Reading, MA: Addison-Wesley, 1981, 1986.

** DATE, C.J. *Relational Database Writings.* Reading, MA: Addison-Wesley, 1986, 1990.

* GHOSH, SAKTI P. *Data Base Organization for Data Management.* New York, NY: Academic Press, 1977.

GRANT, JOHN. *Logical Introduction to Databases.* San Diego, CA: Harcourt Brace Jovanovich, 1987.

KORTH, HENRY F. AND SILBERSCHATZ, ABRAHAM. *Database System Concepts.* New York, NY: McGraw-Hill, 1986.

PAPADIMITRIOU, CHRISTOS H. *The Theory of Database Concurrency Control.* Rockville, MD: Computer Science Press, 1986.

PRATT, PHILIP J. AND ADAMSKI, JOSEPH J. *Database Systems: Management and Design.* San Francisco, CA: Boyd and Fraser, 1987.

* ULLMAN, JEFFREY D. *Principals of Database and Knowledge-based Systems,* 2 Vols. Rockville, MD: Computer Science Press, 1988.

22.6 Programming

*** BENTLEY, JON. *More Programming Pearls.* Reading, MA: Addison-Wesley, 1988.

*** BENTLEY, JON. *Programming Pearls.* Reading, MA: Addison-Wesley, 1986.

* BRAWER, STEVEN. *Introduction to Parallel Programming.* New York, NY: Academic Press, 1989.

** COX, BRAD J. *Object-Oriented Programming: An Evolutionary Approach.* Reading, MA: Addison-Wesley, 1986.

DAHL, O.J.; DIJKSTRA, EDSGER W.; AND HOARE, C.A.R. *Structured Programming.* New York, NY: Academic Press, 1972.

* DIJKSTRA, EDSGER W. *A Discipline of Programming.* Englewood Cliffs, NJ: Prentice Hall, 1976.

FEUER, A. AND GEHANI, NARAIN, EDS. *Comparing and Assessing Programming Languages: Ada, C, Pascal.* Englewood Cliffs, NJ: Prentice Hall, 1984.

HEHNER, ERIC C.R. *The Logic of Programming.* Englewood Cliffs, NJ: Prentice Hall, 1984.

HOGGER, CHRISTOPHER J. *Introduction to Logic Programming.* New York, NY: Academic Press, 1984.

** HOROWITZ, ELLIS, ED. *Programming Languages: A Grand Tour,* Third Edition. Rockville, MD: Computer Science Press, 1983, 1987.

** HOROWITZ, ELLIS. *Fundamentals of Programming Languages,* Second Edition. Rockville, MD: Computer Science Press, 1983, 1984.

*** KERNIGHAN, BRIAN W. AND PLAUGER, P.J. *The Elements of Programming Style,* Second Edition. New York, NY: McGraw-Hill, 1974, 1978.

* MACLENNAN, BRUCE J. *Functional Programming.* Reading, MA: Addison-Wesley, 1989.

MACLENNAN, BRUCE J. *Principles of Programming Languages: Design, Evaluation, and Implementation,* Second Edition. New York, NY: Holt, Rinehart and Winston, 1983, 1986.

MEYER, BERTRAND. *Introduction to the Theory of Programming Languages.* Englewood Cliffs, NJ: Prentice Hall, 1990.

** MEYER, BERTRAND. *Object-Oriented Software Construction,* Second Edition. Englewood Cliffs, NJ: Prentice Hall, 1988, 1991.

PINSON, LEWIS J. AND WIENER, RICHARD S., EDS. *Applications of Object-Oriented Programming.* Reading, MA: Addison-Wesley, 1990.

* READE, CHRIS. *Elements of Functional Programming.* Reading, MA: Addison-Wesley, 1989.

TUCKER, ALLEN B. *Programming Languages,* Second Edition. New York, NY: McGraw-Hill, 1986.

*** WEXELBLAT, RICHARD L., ED. *History of Programming Languages.* New York, NY: Academic Press, 1981.

WIRTH, NIKLAUS. *Systematic Programming: An Introduction.* Englewood Cliffs, NJ: Prentice Hall, 1973.

22.7 Programming Languages

** CLOCKSIN, W.F. AND MELLISH, C.S. *Programming in Prolog,* Third Revised Edition. New York, NY: Springer-Verlag, 1981, 1987.

* COOPER, D. *Standard Pascal: User Reference Manual.* New York, NY: W.W. Norton, 1983.

DEWHURST, STEPHEN C. AND STARK, KATHY T. *Programming in C++.* Englewood Cliffs, NJ: Prentice Hall, 1989.

* GEHANI, NARAIN. *C: An Advanced Introduction.* Rockville, MD: Computer Science Press, 1985, 1988.

** GOLDBERG, ADELE AND ROBSON, DAVID. *Smalltalk-80: The Language.* Reading, MA: Addison-Wesley, 1983, 1989.

HARBISON, SAMUEL P. AND STEELE, GUY L., JR. *C: A Reference Manual,* Third Edition. Englewood Cliffs, NJ: Prentice Hall, 1991.

* JENSEN, KATHLEEN AND WIRTH, NIKLAUS. *Pascal User Manual and Report,* Third Edition. New York, NY: Springer-Verlag, 1985.

*** KERNIGHAN, BRIAN W. AND RITCHIE, DENNIS M. *The C Programming Language,* Second Edition. Englewood Cliffs, NJ: Prentice Hall, 1988.

KOFFMAN, ELLIOT B. *Pascal: Problem Solving and Program Design,* Third Edition. Reading, MA: Addison-Wesley, 1989.

LALONDE, WILF R. AND PUGH, JOHN R. *Inside Smalltalk,* 2 Vols. Englewood Cliffs, NJ: Prentice Hall, 1990.

O'KEEFE, RICHARD A. *The Craft of Prolog.* Cambridge, MA: MIT Press, 1990.

* POLIVKA, RAYMOND AND PAKIN, SANDRA. *APL: The Language and Its Usage.* Englewood Cliffs, NJ: Prentice Hall, 1975.

SEWELL, WAYNE. *Weaving a Program: Literate Programming in WEB.* New York, NY: Van Nostrand Reinhold, 1989.

SHUMATE, K. *Understanding Ada with Abstract Data Types.* New York, NY: John Wiley, 1989.

** STROUSTRUP, BJARNE. *The C++ Programming Language.* Reading, MA: Addison-Wesley, 1987.

WIENER, RICHARD S. AND SINCOVEC, R. *Programming in ADA.* New York, NY: John Wiley, 1983.

** WINSTON, PATRICK H.; KLAUS, BERTHOLD; AND HORN, PAUL. *LISP,* Third Edition. Reading, MA: Addison-Wesley, 1981, 1989.

* WIRTH, NIKLAUS. *Programming in Modula-2,* Fourth Edition. New York, NY: Springer-Verlag, 1982, 1988.

22.8 Algorithms

*** AHO, ALFRED V.; HOPCROFT, JOHN E.; AND ULLMAN, JEFFREY D. *The Design and Analysis of Computer Algorithms.* Reading, MA: Addison-Wesley, 1974.

* BAASE, SARA. *Computer Algorithms: Introduction to Design and Analysis,* Second Edition. Reading, MA: Addison-Wesley, 1978, 1988.

BERLIOUX, PIERRE AND BIZARD, PHILIPPE. *Algorithms: The Construction, Proof, and Analysis of Programs.* New York, NY: John Wiley, 1986.

* BRASSARD, GILLES AND BRATLEY, PAUL. *Algorithmics: Theory and Practice.* Englewood Cliffs, NJ: Prentice Hall, 1988.

BROWN, MARC H. *Algorithm Animation.* Cambridge, MA: MIT Press, 1988.

CORMEN, THOMAS H.; LEISERSON, CHARLES E.; AND RIVEST, RONALD L. *Introduction to Algorithms.* Cambridge, MA: MIT Press, 1990.

* DAVENPORT, J.H.; SIRET, Y.; AND TOURNIER, E. *Computer Algebra: Systems and Algorithms for Algebraic Computation.* New York, NY: Academic Press, 1988.

* DIJKSTRA, EDSGER W. AND FEIJEN, W.H.J. *A Method of Programming.* Reading, MA: Addison-Wesley, 1988.

* GALLIVAN, K.A., *et al.* *Parallel Algorithms for Matrix Computations.* Philadelphia, PA: Society for Industrial and Applied Mathematics, 1990.

* HOROWITZ, ELLIS AND SAHNI, SARTAJ. *Fundamentals of Computer Algorithms.* Rockville, MD: Computer Science Press, 1978.

KEMP, RAINER. *Fundamentals of the Average Case Analysis of Particular Algorithms.* New York, NY: John Wiley, 1984.

KINGSTON, JEFFREY H. *Algorithms and Data Structures: Design, Correctness, Analysis.* Reading, MA: Addison-Wesley, 1990.

*** KNUTH, DONALD E. *The Art of Computer Programming,* 3 Vols., Second Edition. Reading, MA: Addison-Wesley, 1969–81.

** MANBER, UDI. *Introduction to Algorithms: A Creative Approach.* Reading, MA: Addison-Wesley, 1989.

Moret, B.M.E. and Shapiro, H.D. *Algorithms from P to NP.* Redwood City, CA: Benjamin Cummings, 1991.

* Nijenhuis, Albert and Wilf, Herbert S. *Combinatorial Algorithms for Computers and Calculators,* Second Edition. New York, NY: Academic Press, 1975, 1978.

** Reingold, Edward M.; Nievergelt, Jurg; and Deo, Narsingh. *Combinatorial Algorithms: Theory and Practice.* Englewood Cliffs, NJ: Prentice Hall, 1977.

Reisig, Wolfgang. *Petri Nets.* New York, NY: Springer-Verlag, 1985.

** Sedgewick, Robert. *Algorithms in C.* Reading, MA: Addison-Wesley, 1990.

Swartzlander, Earl E., Jr., ed. *Computer Arithmetic.* New York, NY: Dowden, Hutchinson and Ross, 1980.

Wilf, Herbert S. *Algorithms and Complexity.* Englewood Cliffs, NJ: Prentice Hall, 1986.

* Wilf, Herbert S. *Combinatorial Algorithms: An Update.* Philadelphia, PA: Society for Industrial and Applied Mathematics, 1989.

22.9 Theory of Computation

* Brookshear, J. Glenn. *Theory of Computation: Formal Languages, Automata, and Complexity.* Redwood City, CA: Benjamin Cummings, 1989.

Calude, Cristian. *Theories of Computational Complexity.* Amsterdam: North-Holland, 1988.

Davis, Martin D. and Weyuker, Elaine J. *Computability, Complexity, and Languages: Fundamentals of Theoretical Computer Science.* New York, NY: Academic Press, 1983.

*** Garey, Michael R. and Johnson, David S. *Computers and Intractability: A Guide to the Theory of NP-Completeness.* New York, NY: W.H. Freeman, 1979.

* Harrison, Michael A. *Introduction to Formal Language Theory.* Reading, MA: Addison-Wesley, 1978.

* Hartmanis, Juris, ed. *Computational Complexity Theory.* Providence, RI: American Mathematical Society, 1989.

* Moll, Robert N.; Arbib, Michael A.; and Kfoury, A.J. *An Introduction to Formal Language Theory.* New York, NY: Springer-Verlag, 1988.

Salomaa, Arto. *Jewels of Formal Language Theory.* Rockville, MD: Computer Science Press, 1981.

Sanchis, Luis E. *Reflexive Structures: An Introduction to Computability Theory.* New York, NY: Springer-Verlag, 1988.

Winograd, Shmuel. *Arithmetic Complexity of Computations.* Philadelphia, PA: Society for Industrial and Applied Mathematics, 1980.

22.10 Software Systems

Aho, Alfred V.; Kernighan, Brian W.; and Weinberger, Peter J. *The AWK Programming Language.* Reading, MA: Addison-Wesley, 1988.

* André, F.; Herman, D.; and Verjus, J.-P. *Synchronization of Parallel Programs.* Cambridge, MA: MIT Press, 1985.

Beck, Leland L. *System Software,* Second Edition. Reading, MA: Addison-Wesley, 1990.

* Ben-Ari, M. *Principles of Concurrent Programming.* Englewood Cliffs, NJ: Prentice Hall, 1982, 1990.

Chambers, Fred B.; Duce, David A.; and Jones, Gillian P., eds. *Distributed Computing.* New York, NY: Academic Press, 1984.

** Deitel, Harvey M. *Operating Systems,* Second Edition. Reading, MA: Addison-Wesley, 1984, 1990.

Hoare, C.A.R. *Communicating Sequential Processes.* Englewood Cliffs, NJ: Prentice Hall, 1985.

Jones, Oliver. *The X Window System.* Bedford, MA: Digital Press, 1989.

Katzan, Harry. *Operating Systems: A Pragmatic Approach,* Second Edition. New York, NY: Van Nostrand Reinhold, 1986.

* Kernighan, Brian W. and Pike, Rob. *The UNIX Programming Environment.* Englewood Cliffs, NJ: Prentice Hall, 1984.

Krakowiak, Sacha. *Principles of Operating Systems.* Cambridge, MA: MIT Press, 1988.

LEFFLER, SAMUEL J., *et al.* *The Design and Implementation of the 4.3BSD UNIX Operating System.* Reading, MA: Addison-Wesley, 1989.

MILENKOVIC, MILAN. *Operating Systems: Concepts and Design.* New York, NY: McGraw-Hill, 1987.

* SILBERSCHATZ, ABRAHAM; PETERSON, JAMES L.; AND GALVIN, PETER B. *Operating System Concepts,* Third Edition. Reading, MA: Addison-Wesley, 1983, 1991.

* TANENBAUM, ANDREW S. *Operating Systems: Design and Implementation.* Englewood Cliffs, NJ: Prentice Hall, 1987.

22.11 Artificial Intelligence

ALLEN, JAMES. *Natural Language Understanding.* Redwood City, CA: Benjamin Cummings, 1987.

ALTY, J.L. AND COOMBS, M.J. *Expert Systems.* New York, NY: John Wiley, 1984.

* ARBIB, MICHAEL A. *Brains, Machines, and Mathematics,* Second Edition. New York, NY: McGraw-Hill, 1964; New York, NY: Springer-Verlag, 1987.

** BARR, AVRON; COHEN, PAUL R.; AND FEIGENBAUM, EDWARD A., EDS. *The Handbook of Artificial Intelligence,* 4 Vols. Los Altos, CA: William Kaufmann, 1981–82; Reading, MA: Addison-Wesley, 1989.

* BATCHELOR, BRUCE G. *Practical Approach to Pattern Classification.* New York, NY: Plenum Press, 1974.

BOLC, L. AND COOMBS, M.J., EDS. *Expert System Applications.* New York, NY: Springer-Verlag, 1988.

* BUNDY, ALAN. *The Computer Modelling of Mathematical Reasoning.* New York, NY: Academic Press, 1983.

CHOU, SHANG-CHING. *Mechanical Geometry Theorem Proving.* Norwell, MA: D. Reidel, 1988.

FISCHLER, MARTIN A. AND FIRSCHEIN, OSCAR. *Intelligence: The Eye, the Brain, and the Computer.* Reading, MA: Addison-Wesley, 1987.

* FUKUNAGA, KEINOSUKE. *Introduction to Statistical Pattern Recognition,* Second Edition. New York, NY: Academic Press, 1990.

GÖRANZON, BO AND JOSEFSON, INGELA, EDS. *Knowledge, Skill, and Artificial Intelligence.* New York, NY: Springer-Verlag, 1988.

* GRIFFITHS, MICHAEL AND PALISSIER, CAROL. *Algorithmic Methods for Artificial Intelligence.* New York, NY: Chapman and Hall, 1987.

HARRIS, MARY DEE. *Introduction to Natural Language Processing.* Englewood Cliffs, NJ: Reston, 1985.

KANAL, LAVEEN AND KUMAR, VIPIN, EDS. *Search in Artificial Intelligence.* New York, NY: Springer-Verlag, 1988.

KLAHR, PHILIP AND WATERMAN, DONALD A. *Expert Systems.* Reading, MA: Addison-Wesley, 1986.

* KOSKO, BART. *Neural Networks and Fuzzy Systems.* Englewood Cliffs, NJ: Prentice Hall, 1991.

** LEVY, DAVID N.L., ED. *Computer Games I.* New York, NY: Springer-Verlag, 1988.

LUGER, GEORGE F. AND STUBBLEFIELD, WILLIAM A. *Artificial Intelligence and the Design of Expert Systems.* Redwood City, CA: Benjamin Cummings, 1989.

** MINSKY, MARVIN AND PAPERT, SEYMOUR. *Perceptrons: An Introduction to Computational Geometry,* Expanded Edition. Cambridge, MA: MIT Press, 1988.

RAMSAY, ALLAN. *Formal Methods in Artificial Intelligence.* New York, NY: Cambridge University Press, 1988.

ROCH, ELAINE. *Artificial Intelligence.* New York, NY: McGraw-Hill, 1991.

* WIENER, NORBERT. *Cybernetics: Control and Communication in the Animal and the Machine.* New York, NY: John Wiley, 1949.

* WINSTON, PATRICK H. AND BROWN, RICHARD H., EDS. *Artificial Intelligence: An MIT Perspective,* 2 Vols., Second Edition. Cambridge, MA: MIT Press, 1979, 1984.

*** WINSTON, PATRICK H. *Artificial Intelligence.* Reading, MA: Addison-Wesley, 1977.

22.12 Communications

* COLE, ROBERT. *Computer Communications*, Second Edition. New York, NY: Springer-Verlag, 1987.

** COMER, DOUGLAS. *Internetworking with TCP/IP*. Englewood Cliffs, NJ: Prentice Hall, 1988.

JENNINGS, F. *Practical Data Communications: Modems, Networks, and Protocols*. Boston, MA: Blackwell Science, 1986.

* STALLINGS, WILLIAM. *Tutorial: Computer Communications, Architectures, Protocols, and Standards*. Los Angeles, CA: IEEE Computer Society, 1985.

22.13 Mathematical Typography

ABRAHAMS, PAUL W. *TEX for the Impatient*. Reading, MA: Addison-Wesley, 1990.

BUERGER, DAVID J. LATEX *for Scientists and Engineers*. New York, NY: McGraw-Hill, 1990.

DOOB, MICHAEL. *A Gentle Introduction to TEX*. Providence, RI: TEX Users Group, 1990.

KNUTH, DONALD E. *The METAFONTbook*. Reading, MA: Addison-Wesley, 1986.

** KNUTH, DONALD E. *The TEX Book*. Reading, MA: Addison-Wesley, 1986.

* SPIVAK, MICHAEL D. *The Joy of TEX: A Gourmet Guide to Typesetting with the AMS-TEX Macro Package*, Second Edition. Providence, RI: American Mathematical Society, 1986, 1990.

22.14 Software Engineering

* BOOCH, GRADY. *Software Engineering with Ada*, Second Edition. Redwood City, CA: Benjamin Cummings, 1983, 1987.

BROWN, C. MARLIN. *Human-Computer Interface Design Guidelines*. Norwood, NJ: Ablex, 1988.

BROWN, JUDITH R. AND CUNNINGHAM, STEVE. *Programming the User Interface*. New York, NY: John Wiley, 1989.

* GRIES, DAVID. *The Science of Programming*. New York, NY: Springer-Verlag, 1981.

LISKOV, BARBARA AND GUTTAG, JOHN. *Abstraction and Specification in Program Development*. Cambridge, MA: MIT Press, 1986.

* PRESSMAN, R. *Software Engineering: A Practitioner's Approach*. New York, NY: McGraw-Hill, 1987.

SHNEIDERMAN, BEN. *Designing the User Interface*. Reading, MA: Addison-Wesley, 1987.

* SOMMERVILLE, IAN. *Software Engineering*, Second Edition. Reading, MA: Addison-Wesley, 1985.

WELLS, TIMOTHY D. *A Structured Approach to Building Programs: Pascal*. Englewood Cliffs, NJ: Yourdon Press, 1987.

22.15 Automata

BAVEL, ZAMIR. *Introduction to the Theory of Automata*. Englewood Cliffs, NJ: Reston, 1983.

CODD, E.F. *Cellular Automata*. New York, NY: Academic Press, 1968.

* EILENBERG, SAMUEL. *Automata, Languages, and Machines*, 2 Vols. New York, NY: Academic Press, 1974, 1976.

*** HOPCROFT, JOHN E. AND ULLMAN, JEFFREY D. *Introduction to Automata Theory, Languages, and Computation*. Reading, MA: Addison-Wesley, 1979.

* MINSKY, MARVIN. *Computation: Finite and Infinite Machines*. Englewood Cliffs, NJ: Prentice Hall, 1967.

SALOMAA, ARTO. *Computation and Automata*. New York, NY: Cambridge University Press, 1985.

* SAVITCH, WALTER J. *Abstract Machines and Grammars*. Waltham, MA: Little, Brown, 1982.

TOFFOLI, TOMMASO AND MARGOLUS, NORMAN. *Cellular Automata Machines: A New Environment for Modeling*. Cambridge, MA: MIT Press, 1987.

22.16 Compilers and Translators

*** AHO, ALFRED V.; SETHI, RAVI; AND ULLMAN, JEFFREY D. *Compilers: Principles, Techniques, and Tools.* (Former title: *Principles of Compiler Design.*) Reading, MA: Addison-Wesley, 1978, 1986.

* CALINGAERT, PETER. *Program Translation Fundamentals: Methods and Issues.* Rockville, MD: Computer Science Press, 1988.

DAVIE, A.J.T. *Recursive Descent Compiling.* New York, NY: Halsted Press, 1981.

*** FISCHER, CHARLES N. AND LEBLANC, RICHARD J., JR. *Crafting a Compiler.* Redwood City, CA: Benjamin Cummings, 1988.

TERRY, PATRICK D. *Programming Language Translation: A Practical Approach.* Reading, MA: Addison-Wesley, 1987.

** TREMBLAY, JEAN-PAUL. *Theory and Practice of Compiler Writing.* New York, NY: McGraw-Hill, 1985.

WAITE, WILLIAM M. AND GOOS, GERHARD. *Compiler Construction.* New York, NY: Springer-Verlag, 1984.

ZARRELLA, JOHN. *Language Translators.* Suisun City, CA: Microcomputer Applications, 1982.

22.17 Computer Architecture

* BOOTH, TAYLOR L. *Digital Networks and Computer Systems,* Second Edition. New York, NY: John Wiley, 1971, 1978.

CHINITZ, M. PAUL. *The Logic Design of Computers: An Introduction.* Indianapolis, IN: Howard W. Sams, 1981.

* EVANS, DAVID J., ED. *Parallel Processing Systems.* New York, NY: Cambridge University Press, 1982.

FOSTER, CAXTON C. *Computer Architecture,* Third Edition. New York, NY: Van Nostrand Reinhold, 1976, 1985.

FRIEDMAN, ARTHUR D. *Fundamentals of Logic Design and Switching Theory.* Rockville, MD: Computer Science Press, 1986.

GORSLINE, G.W. *Computer Organization: Hardware/Software,* Second Edition. Englewood Cliffs, NJ: Prentice Hall, 1986.

* HILL, FREDRICK J. AND PETERSON, GERALD R. *Digital Systems: Hardware Organization and Design,* Second Edition. New York, NY: John Wiley, 1978.

* LAZOU, CHRISTOPHER. *Supercomputers and their Use,* Revised Edition. New York, NY: Clarendon Press, 1988.

SHIVA, SAJJAN G. *Computer Design and Architecture.* Waltham, MA: Little, Brown, 1985.

* SIEWIOREK, DANIEL P.; BELL, C. GORDON; AND NEWELL, ALLEN. *Computer Structures: Principles and Examples.* New York, NY: McGraw-Hill, 1982.

** STALLINGS, WILLIAM. *Computer Organization and Architecture: Principles of Structure and Function.* New York, NY: Macmillan, 1987.

*** TANENBAUM, ANDREW S. *Structured Computer Organization,* Third Edition. Englewood Cliffs, NJ: Prentice Hall, 1976, 1989.

** WARD, STEPHEN A. AND HALSTEAD, ROBERT H., JR. *Computation Structures.* Cambridge, MA: MIT Press, 1990.

WILLIAMS, CHARLES S. *Designing Digital Filters.* Englewood Cliffs, NJ: Prentice Hall, 1986.

22.18 Computer Graphics

BARTELS, RICHARD H.; BEATTY, JOHN C.; AND BARSKY, BRIAN A. *An Introduction to Splines for Use in Computer Graphics and Geometric Modeling.* San Mateo, CA: Morgan Kaufmann, 1987.

ENDERLE, G.; KANSY, K.; AND PFAFF, G. *Computer Graphics Programming: GKS—The Graphics Standard,* Second Revised Edition. New York, NY: Springer-Verlag, 1984, 1987.

** FIUME, EUGENE L. *The Mathematical Structure of Raster Graphics.* New York, NY: Academic Press, 1989.

*** FOLEY, JAMES D., *et al.* *Computer Graphics: Principles and Practice.* Reading, MA: Addison-Wesley, 1990.

FRIEDHOFF, RICHARD M. AND BENZON, WILLIAM. *The Second Computer Revolution: Visualization.* New York, NY: W.H. Freeman, 1989.

* GLASSNER, ANDREW, ED. *Graphics Gems.* New York, NY: Academic Press, 1990.

** GLASSNER, ANDREW. *An Introduction to Ray Tracing.* New York, NY: Academic Press, 1989.

* HEARN, DONALD AND BAKER, PAULINE. *Computer Graphics.* Englewood Cliffs, NJ: Prentice Hall, 1986.

HÉGRON, GÉRARD. *Image Synthesis: Elementary Algorithms.* Cambridge, MA: MIT Press, 1988.

HOPGOOD, F.R.A., *et al.* *Introduction to the Graphical Kernel System (GKS),* Second Edition. New York, NY: Academic Press, 1986.

* NIELSON, GREGORY M., *et al.* *Visualization in Scientific Computing.* Los Angeles, CA: IEEE Computer Society, 1990.

* PAVLIDIS, THEO. *Algorithms for Graphics and Image Processing.* Rockville, MD: Computer Science Press, 1982.

POKORNY, CORNEL K. AND GERALD, CURTIS F. *Computer Graphics: The Principles Behind the Art and Science.* Irvine, CA: Franklin, Beedle and Associates, 1989.

ROGERS, DAVID F. AND ADAMS, J. ALAN. *Mathematical Elements for Computer Graphics,* Second Edition. New York, NY: McGraw-Hill, 1976, 1990.

** WATT, ALAN. *Fundamentals of Three Dimensional Computer Graphics.* Reading, MA: Addison-Wesley, 1990.

23 Applications to Life Sciences

23.1 General

BATSCHELET, EDWARD. *Introduction to Mathematics for Life Sciences,* Second Edition. New York, NY: Springer-Verlag, 1976.

* BERG, H.C. *Random Walks in Biology.* Princeton, NJ: Princeton University Press, 1983.

BURTON, T.A., ED. *Modeling and Differential Equations in Biology.* New York, NY: Marcel Dekker, 1980.

** CASTI, JOHN L. *Alternate Realities: Mathematical Models of Nature and Man.* New York, NY: John Wiley, 1989.

COOK, THEODORE A. *The Curves of Life.* Mineola, NY: Dover, 1979.

* CULLEN, MICHAEL R. *Mathematics for the Biosciences.* Boston, MA: Prindle, Weber and Schmidt, 1983.

*** EDELSTEIN-KESHET, LEAH. *Mathematical Models in Biology.* Cambridge, MA: Random House, 1988.

GOEL, NARENDA S. AND RICHTER-DYN, NIRA. *Stochastic Models in Biology.* New York, NY: Academic Press, 1974.

** LEVIN, SIMON A., ED. *Studies in Mathematical Biology,* 2 Vols. Washington, DC: Mathematical Association of America, 1978.

* LOTKA, A.J. *Elements of Mathematical Biology.* Mineola, NY: Dover, 1956.

** MARCUS-ROBERTS, HELEN AND THOMPSON, MAYNARD, EDS. *Life Science Models.* Modules in Applied Mathematics, Vol. 4. New York, NY: Springer-Verlag, 1983.

*** MURRAY, J.D. *Mathematical Biology.* New York, NY: Springer-Verlag, 1989.

* OLIVEIRA-PINTO, F. AND CONOLLY, B.W. *Applicable Mathematics of Non-Physical Phenomena.* New York, NY: Halsted Press, 1982.

* RUBINOW, S.I. *Introduction to Mathematical Biology.* New York, NY: John Wiley, 1975.

SMITH, J. MAYNARD. *Mathematical Ideas in Biology.* New York, NY: Cambridge University Press, 1968.

SOLOMON, D.L. AND WALTER, C., EDS. *Mathematical Models in Biological Discovery.* New York, NY: Springer-Verlag, 1977.

** TANUR, JUDITH M., *et al.*, EDS. *Statistics: A Guide to the Study of the Biological and Health Sciences.* San Francisco, CA: Holden-Day, 1977.

23.2 Ecology

BATSCHELET, EDWARD. *Circular Statistics in Biology.* New York, NY: Academic Press, 1981.

* COHEN, JOEL E. *Food Webs and Niche Space.* Princeton, NJ: Princeton University Press, 1978.

DEANGELIS, D.L.; POST, W.M.; AND TRAVIS, C.C. *Positive Feedback in Natural Systems.* New York, NY: Springer-Verlag, 1986.

* DIAMOND, J.M. AND CASE, T.J., EDS. *Community Ecology.* New York, NY: Harper and Row, 1986.

* DUNN, G. AND EVERITT, B.S. *An Introduction to Mathematical Taxonomy.* New York, NY: Cambridge University Press, 1982.

** FREEDMAN, H.I. *Deterministic Mathematical Models in Population Ecology.* New York, NY: Marcel Dekker, 1980.

* GAUSE, G.F. *The Struggle for Existence.* Baltimore, MD: Williams and Wilkins, 1934; Mineola, NY: Dover, 1971.

GITTINS, R. *Canonical Analysis: A Review with Applications in Ecology.* New York, NY: Springer-Verlag, 1985.

** HALLAM, THOMAS G. AND LEVIN, SIMON A., EDS. *Mathematical Ecology: An Introduction.* New York, NY: Springer-Verlag, 1986.

KINGSLAND, SHARON E. *Modeling Nature: Episodes in the History of Population Ecology.* Chicago, IL: University of Chicago Press, 1985.

** LEVIN, SIMON A.; HALLAM, THOMAS G.; AND GROSS, LOUIS J., EDS. *Applied Mathematical Ecology.* New York, NY: Springer-Verlag, 1989.

MACARTHUR, R.H. AND WILSON, EDWARD O. *The Theory of Island Biogeography.* Princeton, NJ: Princeton University Press, 1967.

** MANGEL, MARC AND CLARK, COLIN W. *Dynamic Modeling in Behavioral Ecology.* Princeton, NJ: Princeton University Press, 1988.

* MAY, ROBERT M., ED. *Theoretical Ecology: Principles and Applications.* Boston, MA: Blackwell Science, 1976.

MAY, ROBERT M. *Stability and Complexity in Model Ecosystems,* Second Edition. Princeton, NJ: Princeton University Press, 1974.

MCKELVEY, ROBERT W., ED. *Environmental and Natural Resource Mathematics.* Providence, RI: American Mathematical Society, 1985.

NOBEL, P. *Biophysical Plant Physiology and Ecology.* New York, NY: W.H. Freeman, 1983.

*** OKUBO, A. *Diffusion and Ecological Problems: Mathematical Models.* New York, NY: Springer-Verlag, 1980.

OSTER, GEORGE F. AND WILSON, EDWARD O. *Caste and Ecology in the Social Insects.* Princeton, NJ: Princeton University Press, 1978.

PEDLEY, T.J., ED. *Scale Effects in Animal Locomotion.* New York, NY: Academic Press, 1977.

** PIELOU, E.C. *An Introduction to Mathematical Ecology,* Second Edition. New York, NY: John Wiley, 1969, 1977.

PRUSINKIEWICZ, PRZEMYSLAW AND HANAN, JAMES. *Lindenmayer Systems, Fractals, and Plants.* New York, NY: Springer-Verlag, 1989.

ROUGHGARDEN, JONATHAN; MAY, ROBERT M.; AND LEVIN, SIMON A., EDS. *Perspectives in Ecological Theory.* Princeton, NJ: Princeton University Press, 1989.

* ROUGHGARDEN, JONATHAN. *Theory of Population Genetics and Ecology: An Introduction.* New York, NY: Macmillan, 1979.

SCUDO, F. AND ZIEGLER, J. *The Golden Age of Theoretical Ecology.* New York, NY: Springer-Verlag, 1978.

* SMITH, J. MAYNARD. *Models in Ecology.* New York, NY: Cambridge University Press, 1978.

TILMAN, D. *Resource Competition and Community Structures.* Princeton, NJ: Princeton University Press, 1982.

VINCENT, THOMAS L. AND SKOWRONSKI, JANISLAW M., EDS. *Renewable Resource Management.* New York, NY: Springer-Verlag, 1981.

* VOLTERRA, VITO. *Lecons sur la théorie mathématique de la lutte pour la vie.* Paris: Gauthier-Villars, 1931.

WAITE, T.D. AND FREEMAN, N.J. *Mathematics of Environmental Processes.* Lexington, MA: D.C. Heath, 1977.

WHITTAKER, R.H. AND LEVIN, SIMON A. *Niche: Theory and Application.* New York, NY: Dowden, Hutchinson and Ross, 1975.

WHITTAKER, R.H. *Communities and Ecosystems,* Second Edition. New York, NY: Macmillan, 1970, 1975.

23.3 Epidemiology

ANDERSON, R.M. AND MAY, ROBERT M., EDS. *Population Biology of Infectious Diseases.* New York, NY: Springer-Verlag, 1982.

** ANDERSON, R.M., ED. *Population Dynamics of Infectious Diseases: Theory and Applications.* New York, NY: Chapman and Hall, 1982.

** BAILEY, NORMAN T.J. *The Mathematical Theory of Infectious Diseases and Its Applications,* Second Edition. New York, NY: Macmillan, 1975.

* FRAUENTHAL, J.C. *Mathematical Modeling in Epidemiology.* New York, NY: Springer-Verlag, 1980.

KLEINBAUM, D.G.; KUPPER, L.L.; AND MORGENSTERN, H. *Epidemiologic Research: Principles and Quantitative Methods.* Belmont, CA: Lifetime Learning, 1982.

23.4 Genetics

BULMER, M.G. *The Mathematical Theory of Quantitative Genetics.* New York, NY: Clarendon Press, 1980.

CAVALLI-SFORZA, LUIGI LUCA AND FELDMAN, M.W. *Cultural Transmission and Evolution: A Quantitative Approach.* Princeton, NJ: Princeton University Press, 1981.

CHARNOV, E.L. *The Theory of Sex Allocation.* Princeton, NJ: Princeton University Press, 1982.

* CHRISTIANSEN, F.B. AND FENCHEL, T.M. *Theories of Populations in Biological Communities.* New York, NY: Springer-Verlag, 1977.

** CROW, JAMES R. AND KIMURA, MOTO. *An Introduction to Population Genetics Theory.* New York, NY: Harper and Row, 1970.

*** EWENS, W.J. *Mathematical Population Genetics.* New York, NY: Springer-Verlag, 1979.

** FELDMAN, M.W., ED. *Mathematical Evolutionary Theory.* Princeton, NJ: Princeton University Press, 1989.

FINDLEY, A.M.; MCGLYNN, S.P.; AND FINDLEY, G.L. *The Geometry of Genetics.* New York, NY: John Wiley, 1989.

* GALE, J.S. *Theoretical Population Genetics.* London: Unwin Hyman, 1990.

JACQUARD, ALBERT. *The Genetic Structure of Populations.* New York, NY: Springer-Verlag, 1974.

KARLIN, SAMUEL AND LESSARD, SABIN. *Theoretical Studies on Sex Ratio Evolution.* Princeton, NJ: Princeton University Press, 1986.

* MATHER, K. AND JINKS, J.L. *Biometrical Genetics.* New York, NY: Chapman and Hall, 1982.

MICHOD, R.E. AND LEVIN, B.R., EDS. *The Evolution of Sex.* Sunderland, MA: Sinauer Associates, 1988.

PROVINE, W. *The Origins of Theoretical Population Genetics.* Chicago, IL: University of Chicago Press, 1971.

** SMITH, J. MAYNARD. *Evolution and the Theory of Games.* New York, NY: Cambridge University Press, 1982.

* SUZUKI, DAVID T.; GRIFFITHS, ANTHONY J.F.; AND LEWONTIN, RICHARD C. *An Introduction to Genetic Analysis,* Third Edition. New York, NY: W.H. Freeman, 1976, 1986.

23.5 Population Biology

BARTLETT, M.S. AND HIORNS, R.W., EDS. *The Mathematical Theory of the Dynamics of Biological Populations.* New York, NY: Academic Press, 1973.

** BARTLETT, M.S. *Stochastic Population Models in Ecology and Epidemiology.* New York, NY: Methuen, 1960.

BROWN, ROBERT L. *Introduction to the Mathematics of Demography.* Winsted, CT: ACTEX Publications, 1991.

GINZBURG, L.R. AND GOLENBERG, E.M. *Lectures in Theoretical Population Biology.* Englewood Cliffs, NJ: Prentice Hall, 1985.

* HASSELL, MICHAEL P. *The Dynamics of Arthropod Predator-Prey Systems.* Princeton, NJ: Princeton University Press, 1978.

*** HOPPENSTEADT, FRANK C. *Mathematical Methods of Population Biology.* New York, NY: New York University Press, 1977; New York, NY: Cambridge University Press, 1982.

* HOPPENSTEADT, FRANK C. *Mathematical Theories of Populations: Demographics, Genetics, and Epidemics.* Philadelphia, PA: Society for Industrial and Applied Mathematics, 1975.

HUTCHINSON, G.E. *An Introduction to Population Ecology.* New Haven, CT: Yale University Press, 1978.

IMPAGLIAZZO, J. *Deterministic Aspects of Mathematical Demography.* New York, NY: Springer-Verlag, 1985.

*** KEYFITZ, NATHAN AND BEEKMAN, JOHN A. *Demography Through Problems.* New York, NY: Springer-Verlag, 1984.

* KEYFITZ, NATHAN. *Applied Mathematical Demography.* New York, NY: John Wiley, 1977.

KEYFITZ, NATHAN. *Introduction to the Mathematics of Population,* Revised Edition. Reading, MA: Addison-Wesley, 1968, 1977.

LEWIS, E. *Network Models in Population Biology.* New York, NY: Springer-Verlag, 1977.

** LUDWIG, D. *Stochastic Population Theories.* New York, NY: Springer-Verlag, 1974.

MCDONALD, N. *Time Lags in Biological Models.* New York, NY: Springer-Verlag, 1978.

METZ, J.A.J. AND DIEKMANN, O. *The Dynamics of Physiologically Structured Populations.* New York, NY: Springer-Verlag, 1986.

NISBET, R.M. AND GURNEY, W.S.C. *Modelling Fluctuating Populations.* New York, NY: John Wiley, 1982.

* POLLARD, J.H. *Mathematical Models for the Growth of Human Populations.* New York, NY: Cambridge University Press, 1973.

** SMITH, D.P. AND KEYFITZ, NATHAN. *Mathematical Demography.* New York, NY: Springer-Verlag, 1978.

23.6 Physiology

* CARSON, E.R.; COBELLI, C.; AND FINKELSTEIN, L. *The Mathematical Modeling of Metabolic and Endocrine Systems.* New York, NY: John Wiley, 1983.

COLLINS, R. AND VAN DER WERFF, T.J. *Mathematical Models of the Dynamics of the Human Eye.* New York, NY: Springer-Verlag, 1980.

* GLASS, L. AND MACKEY, M.C. *From Clocks to Chaos: The Rhythms of Life.* Princeton, NJ: Princeton University Press, 1988.

HOLMES, MARK H. AND RUBENFELD, LESTER A., EDS. *Mathematical Modeling of the Hearing Process.* New York, NY: Springer-Verlag, 1981.

** LIGHTHILL, MICHAEL JAMES. *Mathematical Biofluiddynamics.* Philadelphia, PA: Society for Industrial and Applied Mathematics, 1975.

*** MAZUMDAR, J. *An Introduction to Mathematical Physiology and Biology.* New York, NY: Cambridge University Press, 1989.

* MCMAHON, T. *Muscles, Reflexes, and Locomotion.* Princeton, NJ: Princeton University Press, 1984.

PAVLIDIS, THEO. *Biological Oscillators: Their Mathematical Analysis.* New York, NY: Academic Press, 1973.

SCOTT, ALWYN C. *Neurophysics.* New York, NY: John Wiley, 1977.

*** SEGEL, LEE A., ED. *Mathematical Models in Molecular and Cellular Biology.* New York, NY: Cambridge University Press, 1980.

** SEGEL, LEE A. *Modeling Dynamic Phenomena in Molecular and Cellular Biology.* New York, NY: Cambridge University Press, 1984.

STROGATZ, STEVEN H. *The Mathematical Structure of the Human Sleep-Wake Cycle.* New York, NY: Springer-Verlag, 1986.

* THOM, RENÉ. *Mathematical Models of Morphogenesis.* New York, NY: Halsted Press, 1983.

THOM, RENÉ. *Structural Stability and Morphogenesis: An Outline of a General Theory of Models.* Redwood City, CA: Benjamin Cummings, 1975.

THORNLEY, J.H.M. *Mathematical Models in Plant Physiology.* New York, NY: Academic Press, 1976.

VOGEL, S. *Life in Moving Fluids—The Physical Biology of Flow.* Princeton, NJ: Princeton University Press, 1983.

* WINFREE, ARTHUR T. *The Geometry of Biological Time,* Second Edition. New York, NY: Springer-Verlag, 1980, 1990.

** WINFREE, ARTHUR T. *The Timing of Biological Clocks.* New York, NY: Scientific American Library, 1987.

23.7 Medicine

*** BAILAR, JOHN C. AND MOSTELLER, FREDERICK, EDS. *Medical Uses of Statistics.* Waltham, MA: New England Journal of Medicine Books, 1986.

BANKS, H.T. *Modeling and Control in the Biomedical Sciences.* New York, NY: Springer-Verlag, 1975.

CHERRUAULT, Y. *Mathematical Modelling in Biomedicine: Optimal Control of Biomedical Systems.* Norwell, MA: D. Reidel, 1985.

EISEN, MARTIN M. *Mathematical Models in Cell Biology and Cancer Chemotherapy.* New York, NY: Springer-Verlag, 1979.

* GLANTZ, STANTON A. *Mathematics for Biomedical Applications.* Berkeley, CA: University of California Press, 1979.

* HERMAN, GABOR T. *Image Reconstruction from Projections: The Fundamentals of Computerized Tomography.* New York, NY: Academic Press, 1980.

INGRAM, D. AND BLOCH, R.F., EDS. *Mathematical Methods in Medicine,* 2 Vols. New York, NY: John Wiley, 1984.

IOSIFESCU, M. AND TAUTU, P. *Stochastic Processes and Applications in Biology and Medicine,* 2 Vols. New York, NY: Springer-Verlag, 1973.

* JACQUEZ, J.A. *Compartmental Analysis in Biology and Medicine,* Second Edition. Ann Arbor, MI: University of Michigan Press, 1985.

MILLER, RUPERT G., JR., *et al.*, EDS. *Biostatistics Casebook.* New York, NY: John Wiley, 1980.

NATTERER, F. *The Mathematics of Computerized Tomography.* New York, NY: John Wiley, 1986.

* SHEPP, LAWRENCE A. *Computed Tomography.* Providence, RI: American Mathematical Society, 1983.

SWAN, GEORGE W. *Applications of Optimal Control Theory in Biomedicine.* New York, NY: Marcel Dekker, 1984.

23.8 Special Topics

CASTI, JOHN L. AND KARLQVIST, ANDERS, EDS. *Complexity, Language, and Life: Mathematical Approaches.* New York, NY: Springer-Verlag, 1986.

* CHILDRESS, STEPHEN. *Mechanics of Swimming and Flying.* New York, NY: Courant Institute of Mathematical Sciences, 1977.

GERSTENHABER, MURRAY, *et al.*, EDS. *Some Mathematical Questions in Biology,* Vols. 1–18. Providence, RI: American Mathematical Society, 1968–91.

HOPPENSTEADT, FRANK C., ED. *Nonlinear Oscillations in Biology.* Providence, RI: American Mathematical Society, 1979.

JÄGER, W.; ROST, H.; AND TAUTU, P. *Biological Growth and Spread: Mathematical Theories and Applications.* New York, NY: Springer-Verlag, 1980.

** JAGERS, P. *Branching Processes with Biological Applications.* New York, NY: John Wiley, 1975.

* MURRAY, J.D. *Lectures on Nonlinear-Differential-Equation Models in Biology.* New York, NY: Clarendon Press, 1977.

PERCUS, J.K. *Combinatorial Methods in Developmental Biology.* New York, NY: Courant Institute of Mathematical Sciences, 1977.

* PESKIN, CHARLES S. *Partial Differential Equations in Biology.* New York, NY: New York University Press, 1976.

RICCIARDI, L.M. *Diffusion Processes and Related Topics in Biology.* New York, NY: Springer-Verlag, 1977.

24 Applications to Physical Sciences

24.1 General

FRIEDMAN, AVNER. *Mathematics in Industrial Problems.* New York, NY: Springer-Verlag, 1988, 1989.

FRIEDMAN, BERNARD. *Lectures on Applications-Oriented Mathematics.* San Francisco, CA: Holden-Day, 1969.

GEROCH, ROBERT. *Mathematical Physics.* Chicago, IL: University of Chicago Press, 1985.

** HOCHSTADT, HARRY. *The Functions of Mathematical Physics.* New York, NY: John Wiley, 1971.

LEDERMANN, WALTER AND VAJDA, STEVEN, EDS. *Analysis.* Handbook of Applicable Mathematics, Volume IV. New York, NY: John Wiley, 1982.

* NOBLE, BEN. *Applications of Undergraduate Mathematics in Engineering.* Washington, DC: Mathematical Association of America, 1967.

PEARSON, CARL E., ED. *Handbook of Applied Mathematics: Selected Results and Methods,* Second Edition. New York, NY: Van Nostrand Reinhold, 1974, 1983.

PÓLYA, GEORGE. *Mathematical Methods in Science.* Washington, DC: Mathematical Association of America, 1977.

* SUTTON, O.G. *Mathematics in Action.* Mineola, NY: Dover, 1966, 1984.

TAUB, A.H., ED. *Studies in Applied Mathematics.* Washington, DC: Mathematical Association of America, 1971.

* TOWNEND, M. STEWART. *Mathematics in Sport.* New York, NY: Halsted Press, 1984.

WALLACE, PHILIP R. *Mathematical Analysis of Physical Problems.* Mineola, NY: Dover, 1984.

24.2 Introductory Texts

BOAS, MARY L. *Mathematical Methods in the Physical Sciences.* New York, NY: John Wiley, 1966.

COCHRAN, JAMES A. *Applied Mathematics: Principles, Techniques, and Applications.* Belmont, CA: Wadsworth, 1982.

DAVIES, GLYN A.O., ED. *Mathematical Methods in Engineering.* New York, NY: John Wiley, 1984.

* FRIEDMAN, BERNARD. *Principles and Techniques of Applied Mathematics.* New York, NY: John Wiley, 1956.

** GREENBERG, MICHAEL D. *Foundations of Applied Mathematics.* Englewood Cliffs, NJ: Prentice Hall, 1978.

KEENER, JAMES P. *Principles of Applied Mathematics: Transformation and Approximation.* Reading, MA: Addison-Wesley, 1988.

* KOVACH, LADIS D. *Advanced Engineering Mathematics.* Reading, MA: Addison-Wesley, 1982.

KREYSZIG, ERWIN. *Advanced Engineering Mathematics,* Fifth Edition. New York, NY: John Wiley, 1972, 1983.

MATHEWS, JON AND WALKER, ROBERT J. *Mathematical Methods of Physics.* Reading, MA: W.A. Benjamin, 1970.

MEYER, RICHARD M. *Essential Mathematics for Applied Fields.* New York, NY: Springer-Verlag, 1979.

* POLLARD, HARRY. *Introduction to Applied Mathematics.* Reading, MA: Addison-Wesley, 1972.

SIROVICH, L. *Introduction to Applied Mathematics.* New York, NY: Springer-Verlag, 1988.

SOKOLNIKOFF, IVAN S. AND REDHEFFER, RAYMOND M. *Mathematics of Physics and Modern Engineering,* Second Edition. New York, NY: McGraw-Hill, 1966.

*** STRANG, GILBERT. *Introduction to Applied Mathematics.* Wellesley, MA: Wellesley-Cambridge Press, 1986.

WYLIE, CLARENCE R. *Advanced Engineering Mathematics,* Third Edition. New York, NY: McGraw-Hill, 1966.

24.3 Advanced Surveys

BENDER, CARL AND ORSZAG, STEVEN. *Advanced Mathematical Methods for Scientists and Engineers.* New York, NY: McGraw-Hill, 1978.

*** COURANT, RICHARD AND HILBERT, DAVID. *Methods of Mathematical Physics,* 2 Vols. New York, NY: John Wiley, 1974, 1989.

* JEFFREYS, SIR HAROLD AND JEFFREYS, BERTHA S. *Methods of Mathematical Physics,* Third Edition. New York, NY: Cambridge University Press, 1956.

** LIN, C.C. AND SEGEL, LEE A. *Mathematics Applied to Deterministic Problems in the Natural Sciences.* New York, NY: Macmillan, 1974.

NOWINSKI, J.L. *Applications of Functional Analysis in Engineering.* New York, NY: Plenum Press, 1981.

* REED, MICHAEL AND SIMON, BARRY. *Methods of Modern Mathematical Physics,* 4 Vols. New York, NY: Academic Press, 1972–79.

STAKGOLD, IVAR. *Boundary Value Problems in Mathematical Physics,* 2 Vols. New York, NY: Macmillan, 1968.

* THOMPSON, J.M.T. *Instabilities and Catastrophes in Science and Engineering.* New York, NY: John Wiley, 1982.

VON WESTENHOLZ, C. *Differential Forms in Mathematical Physics.* Amsterdam: North-Holland, 1978.

ZADEH, LOTFI AND DESOER, C.A. *Linear Systems Theory.* New York, NY: McGraw-Hill, 1963.

24.4 Elasticity and Solid Mechanics

* FUNG, Y.C. *Foundations of Solid Mechanics.* Englewood Cliffs, NJ: Prentice Hall, 1965.

GREEN, A.E. AND ZERNA, W. *Theoretical Elasticity,* Second Edition. New York, NY: Oxford University Press, 1968.

HERTZBERG, RICHARD W. *Deformation and Fracture Mechanics of Engineering Materials,* Third Edition. New York, NY: John Wiley, 1976, 1989.

** HILL, R. *The Mathematical Theory of Plasticity.* New York, NY: Oxford University Press, 1983.

*** LOVE, A.E.H. *Treatise on the Mathematical Theory of Elasticity,* Fourth Revised Edition. Mineola, NY: Dover, 1956.

MEISOVITCH, LEONARD. *Elements of Vibration Analysis,* Second Edition. New York, NY: McGraw-Hill, 1975, 1986.

* SOKOLNIKOFF, IVAN S. *Mathematical Theory of Elasticity,* Second Edition. New York, NY: McGraw-Hill, 1956; Melbourne, FL: Robert E. Krieger, 1983.

** STOKER, J.J. *Nonlinear Vibrations in Mechanical and Electrical Systems.* New York, NY: Interscience, 1950.

* TIMOSHENKO, STEPHEN AND GOODIER, J.N. *Theory of Elasticity,* Second Edition. New York, NY: McGraw-Hill, 1934, 1951.

TIMOSHENKO, STEPHEN AND WOINOWSKY-KRIEGER, S. *Theory of Plates and Shells,* Second Edition. New York, NY: McGraw-Hill, 1959.

TIMOSHENKO, STEPHEN. *Theory of Elastic Stability,* Second Edition. New York, NY: McGraw-Hill, 1961.

24.5 Quantum Mechanics

BETHE, HANS A. AND SALPETER, EDWIN E. *Quantum Mechanics for One- and Two-Electron Atoms.* New York, NY: Plenum Press, 1977.

CHESNUT, D.B. *Finite Groups and Quantum Theory.* Melbourne, FL: Robert E. Krieger, 1982.

* D'ESPAGNAT, BERNARD. *Conceptual Foundations of Quantum Mechanics,* Second Edition. Redwood City, CA: Benjamin Cummings, 1976.

DEWITT, BRYCE S. AND GRAHAM, NEILL, EDS. *The Many-Worlds Interpretation of Quantum Mechanics.* Princeton, NJ: Princeton University Press, 1973.

* GUTZWILLER, MARTIN C. *Chaos in Classical and Quantum Mechanics.* New York, NY: Springer-Verlag, 1990.

JORDAN, THOMAS F. *Quantum Mechanics in Simple Matrix Form.* New York, NY: John Wiley, 1986.

MACKEY, GEORGE W. *The Mathematical Foundations of Quantum Mechanics.* Reading, MA: W.A. Benjamin, 1963.

24.6 Mechanics

ANGELES, JORGE. *Rational Kinematics.* New York, NY: Springer-Verlag, 1988.

** ARNOLD, V.I. *Mathematical Methods of Classical Mechanics,* Second Edition. New York, NY: Springer-Verlag, 1978, 1989.

*** GOLDSTEIN, HERBERT. *Classical Mechanics.* Reading, MA: Addison-Wesley, 1950.

LANCZOS, CORNELIUS. *Variational Principles of Mechanics,* Fourth Edition. Toronto: University of Toronto Press, 1970.

RASBAND, S. NEIL. *Dynamics.* New York, NY: John Wiley, 1983.

24.7 Relativity

CALDER, NIGEL. *Einstein's Universe.* New York, NY: Viking Press, 1979.

CLARKE, C. *Elementary General Relativity.* New York, NY: John Wiley, 1979.

EDDINGTON, ARTHUR S. *The Mathematical Theory of Relativity,* Third Edition. New York, NY: Cambridge University Press, 1924; New York, NY: Chelsea, 1975.

EINSTEIN, ALBERT. *The Meaning of Relativity.* Fourth Edition. Princeton, NJ: Princeton University Press, 1946, 1953.

EINSTEIN, ALBERT. *Relativity: The Special and the General Theory.* London: Methuen, 1920; New York, NY: Crown 1961.

** GEROCH, ROBERT. *General Relativity from A to B.* Chicago, IL: University of Chicago Press, 1978.

* LAWDEN, DEREK F. *Elements of Relativity Theory.* New York, NY: John Wiley, 1985.

* LIEBER, LILLIAN R. *The Einstein Theory of Relativity*; Corte Madera, CA: Rinehart Press, 1945.

*** MISNER, CHARLES W.; THORNE, KIP S.; AND WHEELER, JOHN ARCHIBALD. *Gravitation.* New York, NY: W.H. Freeman, 1973.

* RINDLER, WOLFGANG. *Essential Relativity: Special, General, and Cosmological,* Second Edition. New York, NY: Springer-Verlag, 1977.

24.8 Cosmology and Celestial Mechanics

DAVIES, P.C.W. *Space and Time in the Modern Universe.* New York, NY: Cambridge University Press, 1977.

HOYLE, FRED. *From Stonehenge to Modern Cosmology.* New York, NY: W.H. Freeman, 1972.

* LANDSBERG, PETER L. AND EVANS, DAVID A. *Mathematical Cosmology: An Introduction.* New York, NY: Clarendon Press, 1977.

** POLLARD, HARRY. *Mathematical Introduction to Celestial Mechanics.* Englewood Cliffs, NJ: Prentice Hall, 1966; Washington, DC: Mathematical Association of America, 1976.

** SZEBEHELY, VICTOR G. *Adventures in Celestial Mechanics: A First Course in the Theory of Orbits.* Austin, TX: University of Texas Press, 1989.

WALD, ROBERT M. *Space, Time, and Gravity: The Theory of the Big Bang and Black Holes.* Chicago, IL: University of Chicago Press, 1977.

24.9 Fluids

* ACHESON, D.J. *Elementary Fluid Dynamics.* New York, NY: Oxford University Press, 1990.

*** BATCHELOR, GEORGE K. *Introduction to Fluid Dynamics.* New York, NY: Cambridge University Press, 1967.

CHORIN, A.J. AND MARSDEN, JERROLD E. *A Mathematical Introduction to Fluid Mechanics,* Second Edition. New York, NY: Springer-Verlag, 1990.

JONES, D.S. *Acoustic and Electromagnetic Waves.* New York, NY: Oxford University Press, 1986.

* LAMB, SIR HORACE. *Hydrodynamics,* Sixth Revised Edition. Mineola, NY: Dover, 1956.

LANGLOIS, W.E. *Slow Viscous Flow.* New York, NY: Macmillan, 1964.

** LIGHTHILL, MICHAEL JAMES. *An Informal Introduction to Theoretical Fluid Mechanics.* New York, NY: Oxford University Press, 1986.

MEYER, RICHARD E. *Introduction to Mathematical Fluid Dynamics.* New York, NY: John Wiley, 1971; Mineola, NY: Dover, 1982.

* MILNE-THOMSON, L.M. *Theoretical Hydrodynamics,* Fifth Edition. New York, NY: Macmillan, 1950; New York, NY: St. Martin's Press, 1960.

SEGEL, LEE A. AND HANDELMAN, G.H. *Mathematics Applied to Continuum Mechanics.* New York, NY: Macmillan, 1977.

* STOKER, J.J. *Water Waves: The Mathematical Theory with Applications.* New York, NY: Interscience, 1957.

24.10 Waves

BALDOCK, G.R. AND BRIDGEMAN, T. *Mathematical Theory of Wave Motion.* New York, NY: John Wiley, 1981.

* COURANT, RICHARD AND FRIEDRICHS, KURT OTTO. *Supersonic Flow and Shock Waves.* New York, NY: Springer-Verlag, 1976.

DAVIS, JULIAN L. *Wave Propagation in Solids and Fluids.* New York, NY: Springer-Verlag, 1988.

** LIGHTHILL, MICHAEL JAMES. *Waves in Fluids.* New York, NY: Cambridge University Press, 1978.

MÜLLER, C. *Foundations of the Mathematical Theory of Electromagnetic Waves.* New York, NY: Springer-Verlag, 1969.

* RAYLEIGH, LORD JOHN. *Theory of Sound,* 2 Vols., Second Revised Edition. Gloucester, MA: Peter Smith, 1955.

24.11 Tensor Analysis

* BORISENKO, A.I. AND TARAPOV, I.E. *Vector and Tensor Analysis with Applications.* Mineola, NY: Dover, 1979.

CHORLTON, FRANK. *Vector & Tensor Methods.* New York, NY: John Wiley, 1976.

LAWDEN, DEREK F. *An Introduction to Tensor Calculus, Relativity, and Cosmology,* Third Edition. New York, NY: John Wiley, 1982.

** SOKOLNIKOFF, IVAN S. *Tensor Analysis: Theory and Applications to Geometry and Mechanics of Continua,* Second Edition. New York, NY: John Wiley, 1964.

24.12 Information Theory

CADZOW, JAMES A. *Foundations of Digital Signal Processing and Data Analysis.* New York, NY: Macmillan, 1987.

CATTERMOLE, KENNETH W. AND O'REILLY, JOHN J., EDS. *Problems of Randomness in Communication Engineering.* New York, NY: John Wiley, 1984.

* CHAMBERS, WILLIAM G. *Basics of Communications and Coding.* New York, NY: Oxford University Press, 1985.

** JONES, D.S. *Elementary Information Theory.* New York, NY: Clarendon Press, 1979.

* OBERMAN, R.M.M. *Digital Circuits for Binary Arithmetic.* New York, NY: John Wiley, 1979.

RADER, JAMES H. AND MCCLELLAN, CHARLES M. *Number Theory in Digital Signal Processing.* Englewood Cliffs, NJ: Prentice Hall, 1979.

** RÉNYI, ALFRÉD. *A Diary on Information Theory.* Budapest: Akademiai Kiado, 1984; New York, NY: John Wiley, 1987.

*** SHANNON, CLAUDE E. AND WEAVER, WARREN. *The Mathematical Theory of Communication.* Champaign, IL: University of Illinois Press, 1949.

24.13 Special Topics

BAILEY, T.N. AND BASTON, R.J., EDS. *Twistors in Mathematics and Physics.* New York, NY: Cambridge University Press, 1990.

* BOISEN, M.B., JR. AND GIBBS, G.V. *Mathematical Crystallography.* Washington, DC: Mineralogical Society of America, 1985.

* GHEZ, RICHARD. *A Primer of Diffusion Problems.* New York, NY: John Wiley, 1988.

* HART, DEREK AND CROFT, TONY. *Modelling with Projectiles.* New York, NY: Halsted Press, 1988.

* ISENBERG, CYRIL. *The Science of Soap Films and Soap Bubbles.* Clevedon: Tieto Ltd., 1978.

** OWEN, DAVID R. *A First Course in the Mathematical Foundations of Thermodynamics.* New York, NY: Springer-Verlag, 1984.

ROMANOVA, M.A. AND SARMANOV, O.V., EDS. *Topics in Mathematical Geology.* New York, NY: Plenum Press, 1970.

* THOMPSON, COLIN J. *Mathematical Statistical Mechanics.* Princeton, NJ: Princeton University Press, 1972.

WOOD, ELIZABETH A. *Crystals and Light: An Introduction to Optical Crystallography,* Second Revised Edition. Mineola, NY: Dover, 1977.

25 Applications to Social Sciences

25.1 General

BARTHOLOMEW, DAVID J. *Mathematical Methods in Social Science.* New York, NY: John Wiley, 1981.

* BARTHOLOMEW, DAVID J. *Stochastic Models for Social Processes,* Second Edition. New York, NY: John Wiley, 1973, 1978.

BERLINSKI, DAVID. *On System Analysis: An Essay Concerning the Limitations of Some Mathematical Methods in the Social, Political, and Biological Sciences.* Cambridge, MA: MIT Press, 1976.

BRADLEY, IAN AND MEEK, RONALD L. *Matrices and Society.* Princeton, NJ: Princeton University Press, 1986.

DORAN, J.E. AND HODSON, F.R. *Mathematics and Computers in Archaeology.* Cambridge, MA: Harvard University Press, 1975.

GOLDBERG, SAMUEL I. *Probability in Social Science.* New York, NY: Birkhäuser, 1983.

GOULD, STEPHEN JAY. *The Mismeasure of Man.* New York, NY: W.W. Norton, 1981.

** GROSS, MAURICE. *Mathematical Models in Linguistics.* Englewood Cliffs, NJ: Prentice Hall, 1972.

KEMENY, JOHN G. AND SNELL, J. LAURIE. *Mathematical Models in the Social Sciences.* Cambridge, MA: MIT Press, 1972.

KIM, KI HANG AND ROUSH, FRED W. *Mathematics for Social Scientists.* New York, NY: Elsevier Science, 1980.

* LAZARSFELD, PAUL F., ED. *Mathematical Thinking in the Social Sciences.* New York, NY: Russell and Russell, 1969.

MAKI, DANIEL P. AND THOMPSON, MAYNARD. *Mathematical Models and Applications with Emphasis on the Social, Life, and Management Sciences.* Englewood Cliffs, NJ: Prentice Hall, 1973.

OLINICK, MICHAEL. *An Introduction to Mathematical Models in the Social and Life Sciences.* Reading, MA: Addison-Wesley, 1978.

* RAPOPORT, ANATOL. *Mathematical Models in the Social and Behavioral Sciences.* New York, NY: John Wiley, 1983.

* ROBERTS, FRED S. *Graph Theory and Its Applications to Problems of Society.* Philadelphia, PA: Society for Industrial and Applied Mathematics, 1978.

*** TANUR, JUDITH M., *et al.*, EDS. *Statistics: A Guide to Political and Social Issues.* San Francisco, CA: Holden-Day, 1977.

** WASHBURN, D.K. AND CROWE, DONALD W. *Symmetries of Culture: Handbook of Plane Pattern Analysis.* Seattle, WA: University of Washington Press, 1988.

25.2 Economics

* ARROW, KENNETH J. AND INTRILIGATOR, MICHAEL D., EDS. *Handbook of Mathematical Economics,* 3 Vols. New York, NY: Elsevier Science, 1981–86.

* BALASKO, YVES. *Foundations of the Theory of General Equilibrium.* New York, NY: Academic Press, 1987.

BANERJEE, KALI S. *Cost of Living Index Numbers: Practice, Precision, and Theory.* New York, NY: Marcel Dekker, 1975.

** BOOT, JOHN C.G. *Common Globe or Global Commons: Population Regulation and Income Distribution.* New York, NY: Marcel Dekker, 1974.

* CLARK, COLIN W. *Bioeconomic Modelling and Fisheries Management.* New York, NY: John Wiley, 1985.

*** CLARK, COLIN W. *Mathematical Bioeconomics: The Optimal Management of Renewable Resources,* Second Edition. New York, NY: John Wiley, 1976, 1990.

* DEBREU, GERARD. *Theory of Value: An Axiomatic Analysis of Economic Equilibrium.* New Haven, CT: Yale University Press, 1959, 1971.

FRANKLIN, JOEL. *Methods of Mathematical Economics: Linear and Nonlinear Programming, Fixed-Point Theorems.* New York, NY: Springer-Verlag, 1980.

FRIEDMAN, JAMES W. *Oligopoly and the Theory of Games.* Amsterdam: North-Holland, 1977.

* GALE, DAVID. *The Theory of Linear Economic Models.* New York, NY: McGraw-Hill, 1960.

* GRANDMONT, JEAN-MICHEL. *Temporary Equilibrium: Selected Readings.* New York, NY: Academic Press, 1988.

** GRILICHES, ZVI AND INTRILIGATOR, MICHAEL D., EDS. *Handbook of Econometrics,* 2 Vols. New York, NY: Elsevier Science, 1983.

** INGRAO, BRUNA AND ISRAEL, GIORGIO. *The Invisible Hand: Economic Equilibrium in the History of Science.* Cambridge, MA: MIT Press, 1990.

JAMES, D.E. AND THROSBY, C.D. *Introduction to Quantitative Methods in Economics.* New York, NY: John Wiley, 1974.

* KENNEDY, PETER. *A Guide to Econometrics,* Second Edition. Cambridge, MA: MIT Press, 1985.

KHOURY, SARKIS J. AND PARSONS, TORRENCE D. *Mathematical Methods in Finance and Economics.* New York, NY: Elsevier Science, 1981.

* KLEIN, ERVIN. *Mathematical Methods in Theoretical Economics: Topological and Vector Space Foundations of Equilibrium Analysis.* New York, NY: Academic Press, 1973.

* LEONTIEF, W. *Input-Output Economics.* New York, NY: Oxford University Press, 1966.

MAKAROV, V.L. AND RUBINOV, A.M. *Mathematical Theory of Economic Dynamics and Equilibria.* New York, NY: Springer-Verlag, 1977.

NIKAIDO, HUKUKANE. *Convex Structures and Economic Theory.* New York, NY: Academic Press, 1968.

RADER, TROUT. *Theory of Microeconomics.* New York, NY: Academic Press, 1972.

*** RAIFFA, HOWARD. *The Art and Science of Negotiation.* Cambridge, MA: Harvard University Press, 1982.

** REITER, STANLEY, ED. *Studies in Mathematical Economics.* Washington, DC: Mathematical Association of America, 1986.

SASSONE, PETER G. AND SCHAFFER, WILLIAM A. *Cost-Benefit Analysis: A Handbook.* New York, NY: Academic Press, 1978.

SATO, RYUZO. *Theory of Technical Change and Economic Invariance: Application of Lie Groups.* New York, NY: Academic Press, 1981.

* SCARF, HERBERT. *The Computation of Economic Equilibria.* New Haven, CT: Yale University Press, 1973.

** STARR, ROSS M. *General Equilibrium Models of Monetary Economics: Studies in the Static Foundation of Monetary Theory.* New York, NY: Academic Press, 1989.

* STIGUM, BRENT P. *Toward a Formal Science of Economics: The Axiomatic Method in Economics and Econometrics.* Cambridge, MA: MIT Press, 1990.

VAN DER PLOEG, FREDERICK, ED. *Mathematical Methods in Economics.* New York, NY: John Wiley, 1984.

** WONNACOTT, RONALD J. AND WONNACOTT, THOMAS H. *Econometrics,* Second Edition. New York, NY: John Wiley, 1979.

25.3 Political Science

*** BALINSKI, MICHAEL L. AND YOUNG, H. PEYTON. *Fair Representation: Meeting the Ideal of One Man, One Vote.* New Haven, CT: Yale University Press, 1982.

BELTRAMI, EDWARD J. *Models for Public Systems Analysis.* New York, NY: Academic Press, 1977.

* BRAMS, STEVEN J. AND FISHBURN, PETER C. *Approval Voting.* New York, NY: Birkhäuser, 1983.

** BRAMS, STEVEN J.; LUCAS, WILLIAM F.; AND STRAFFIN, PHILIP D., JR., EDS. *Political and Related Models.* Modules in Applied Mathematics, Vol. 2. New York, NY: Springer-Verlag, 1983.

* BRAMS, STEVEN J. *Paradoxes in Politics: An Introduction to the Nonobvious in Political Science.* New York, NY: Free Press, 1976.

BRAMS, STEVEN J. *Superpower Games: Applying Game Theory to Superpower Conflict.* New Haven, CT: Yale University Press, 1985.

FISHBURN, PETER C. *The Theory of Social Choice.* Princeton, NJ: Princeton University Press, 1973.

KIM, KI HANG AND ROUSH, FRED W. *Introduction to Mathematical Consensus Theory.* New York, NY: Marcel Dekker, 1980.

** MIRKIN, BORIS G. *Group Choice.* Silver Spring, MD: V.H. Winston, 1979.

ORDESHOOK, PETER C., ED. *Game Theory and Political Science.* New York, NY: New York University Press, 1978.

** STRAFFIN, PHILIP D., JR. *Topics in the Theory of Voting.* New York, NY: Birkhäuser, 1980.

25.4 Psychology

ATKINSON, RICHARD C., *et al. Introduction to Mathematical Learning Theory.* New York, NY: John Wiley, 1965.

* LUCE, ROBERT DUNCAN, *et al. Handbook of Mathematical Psychology,* 3 Vols. New York, NY: John Wiley, 1963–65.

MESSICK, DAVID M., ED. *Mathematical Thinking in Behavioral Sciences,* Readings from Scientific American. New York, NY: W.H. Freeman, 1968.

TVERSKY, AMOS; COOMBS, CLYDE H.; AND DAWES, ROBYN M. *Mathematical Psychology: An Elementary Introduction.* Englewood Cliffs, NJ: Prentice Hall, 1970.

25.5 Sociology

BEAUCHAMP, MURRAY A. *Elements of Mathematical Sociology.* Philadelphia, PA: Philadelphia Books, 1970.

** BERGER, JOSEPH, *et al. Types of Formalization in Small Group Research.* Boston, MA: Houghton Mifflin, 1962.

* COLEMAN, JAMES S. *Introduction to Mathematical Sociology.* New York, NY: Free Press, 1964.

* FARARO, THOMAS J. *Mathematical Sociology: An Introduction to Fundamentals.* New York, NY: John Wiley, 1973.

26 Journals and Periodicals

Journals and periodicals listed below are those that include articles of potential interest to undergraduate students in the mathematical sciences. With few exceptions, research journals are not listed. The journals listed below are devoted primarily to the mathematical sciences. In addition, many general science periodicals (such as *Scientific American* and *Sciences News*) include excellent articles about the mathematical sciences and should be part of any good undergraduate library.

AMATYC Review
* AMSTAT News
American Journal of Mathematics
*** American Mathematical Monthly
American Statistician
Archive for the History of the Exact Sciences
* Arithmetic Teacher
Bulletin of the American Mathematical Society
Bulletin of the London Mathematical Society
Byte
COMAP Consortium
** Chance
*** College Mathematics Journal
Collegiate Microcomputer
* Communications of the ACM
* Computer Science Education
Computing Reviews
Computing Surveys of the ACM
Crux Mathematicorum
Current Mathematics Publications
Educational Studies in Mathematics
Elemente der Mathematik
Fibonacci Quarterly
* Historia Mathematica
* Interfaces (TIMS)
International Journal of Mathematics Education in Science and Technology
* Journal for Research in Mathematics Education
Journal of Applied Probability
Journal of Number Theory
* Journal of Recreational Mathematics
Journal of Technology in Mathematics
* Journal of Undergraduate Mathematics
Journal of the American Mathematical Society
Journal of the American Statistical Association
L'Enseignement Mathematique
* Mathematical Gazette
*** Mathematical Intelligencer
** Mathematical Reviews
* Mathematical Spectrum
*** Mathematics Magazine
** Mathematics Teacher
Mathematics and Computer Education
Mathematics of Computation

** Notices of the American Mathematical Society
* OR/MS Today
Operations Research
* Pi Mu Epsilon Journal
Primus
Proceedings of the American Mathematical Society
** Quantum
SIAM Journal on Applied Mathematics
* SIAM News
* SIAM Review
School Science and Mathematics
** Statistical Science (IMS)
* Stats (ASA)
Sugaku Expositions (AMS Translations)
Transactions of the American Mathematical Society
** UMAP Journal
** Undergraduate Mathematics Education (UME) Trends

Index of Authors

A

Amasigo, John C. *Advanced Calculus and Its Applications to the Engineering and Physical Sciences.* [6.4]

Ambartsumian, R.V. *Combinatorial Integral Geometry with Applications to Mathematical Stereology.* [14.9]

* American Association for the Advancement of Science. *Science For All Americans.* [5.6]

*** American Council of Learned Societies. *Biographical Dictionary of Mathematicians.* [2.3]

* American Education Research Association. *Research in Teaching and Learning.* [5.12]

* Anderson, Arthur W. *Pension Mathematics for Actuaries.* [17.8]

* Anderson, Ian. *A First Course in Combinatorial Mathematics.* [10.3]

* ———. *Combinatorial Designs: Construction Methods.* [10.4]

** ———. *Combinatorics of Finite Sets.* [10.4]

Anderson, J.G. *Technical Shop Mathematics.* [16.2]

Anderson, John T. *Excursions in Number Theory.* [11.2]

Anderson, R.M. *Population Biology of Infectious Diseases.* [23.3]

** ———. *Population Dynamics of Infectious Diseases: Theory and Applications.* [23.3]

Anderson, T.W. *A Bibliography of Multivariate Statistical Analysis.* [2.4]

———. *An Introduction to Multivariate Statistical Analysis.* [21.10]

* André, F. *Synchronization of Parallel Programs.* [22.10]

* Andrews, D.F. *Data: A Collection of Problems from Many Fields for the Student and Research Worker.* [21.6]

Andrews, George E. *Actuarial Projections for the Old Age, Survivor, and Disability Income Program of Social Security in the United States of America.* [17.8]

* ———. *Number Theory.* [11.3]

———. *q-Series: Their Development and Applications in Analysis, Number Theory, Combinatorics, Physics, and Computer Algebra.* [8.15]

** ———. *The Theory of Partitions.* [11.9]

* Andrews, J.G. *Mathematical Modelling.* [19.2]

Andrews, Larry C. *Elementary Partial Differential Equations with Boundary Value Problems.* [7.4]

Andrews, Peter B. *An Introduction to Mathematical Logic and Type Theory: To Truth Through Proof.* [9.3]

Andrews, William S. *Magic Squares and Cubes.* [4.2]

Angeles, Jorge. *Rational Kinematics.* [24.6]

* Anton, Howard. *Applied Finite Mathematics.* [10.2]

* ———. *Applications of Linear Algebra.* [12.1]

* ———. *Calculus with Analytic Geometry.* [6.3]

** ———. *Elementary Linear Algebra.* [12.1]

** Apostol, Tom M. *Selected Papers on Calculus.* [6.5]

** ———. *Selected Papers on Precalculus.* [6.5]

*** ———. *Calculus.* [6.4]

*** ———. *Introduction to Analytic Number Theory.* [11.6]

* ———. *Mathematical Analysis.* [8.2]

* ———. *Modular Functions and Dirichlet Series in Number Theory.* [11.7]

ApSimon, Hugh. *Mathematical Byways in Ayling, Beeling, and Ceiling.* [4.2]

* Arbib, Michael A. *An Introduction to Formal Language Theory.* [22.9]

* ———. *Brains, Machines, and Mathematics.* [22.11]

Arganbright, Dean E. *Mathematical Applications of Electronic Spreadsheets.* [22.1]

Arkhangelskiĭ, A.V. *General Topology I: Basic Concepts and Constructions, Dimension Theory.* [15.1]

** Armstrong, M.A. *Basic Topology.* [15.3]

** ———. *Groups and Symmetry.* [14.5]

Arney, William R. *Understanding Statistics in the Social Sciences.* [21.2]

* Arnold, V.I. *Catastrophe Theory.* [14.10]

** ———. *Geometrical Methods in the Theory of Ordinary Differential Equations.* [7.2]

———. *Huygens and Barrow, Newton and Hooke.* [3.9]

** ———. *Mathematical Methods of Classical Mechanics.* [24.6]

* ———. *Ordinary Differential Equations.* [7.2]

* Arrow, Kenneth J. *Handbook of Mathematical Economics.* [25.2]

* Arrowsmith, D.K. *An Introduction to Dynamical Systems.* [7.3]

Artin, Emil. *Class Field Theory.* [11.5]

* ———. *Galois Theory.* [13.7]

———. *Geometric Algebra.* [13.13]

* ———. *Introduction to Algebraic Topology.* [15.3]

** ———. *The Gamma Function.* [8.15]

** Artin, Michael. *Algebra.* [13.1]

Artzy, Rafael. *Linear Geometry.* [14.12]

Arveson, William. *An Invitation to C^*-Algebras.* [8.9]

Aschbacher, Michael. *The Finite Simple Groups and Their Classification.* [13.5]

Ascher, Marcia. *Code of the Quipu: A Study in Media, Mathematics, and Culture.* [5.4]

* ———. *Ethnomathematics: A Multicultural View of Mathematical Ideas.* [5.4]

Ascher, Robert. *Code of the Quipu: A Study in Media, Mathematics, and Culture.* [5.4]

Ash, Carol. *The Calculus Tutoring Book.* [6.3]

———. *Introduction to Discrete Mathematics.* [10.1]

** Ash, J.M. *Studies in Harmonic Analysis.* [8.12]

Ash, Robert B. *Real Analysis and Probability.* [20.6]

———. *The Calculus Tutoring Book.* [6.3]

Asimov, Isaac. *Asimov on Numbers.* [1.2]

** Askey, Richard A. *A Century of Mathematics in America.* [3.6]

Aspray, William. *History and Philosophy of Modern Mathematics.* [9.4]

———. *John von Neumann and the Origins of Modern Computing.* [3.2]

Athen, Hermann. *Proceedings of the Third International Congress on Mathematical Education.* [5.6]

Athreya, A.K.B. *Branching Processes.* [19.3]

** Atiyah, Michael F. *Introduction to Commutative Algebra.* [13.8]

* ———. *The Geometry and Physics of Knots.* [15.2]

B

Bartee, Thomas C. *Modern Applied Algebra.* [13.3]
Bartels, Richard H. *An Introduction to Splines for Use in Computer Graphics and Geometric Modeling.* [22.18]
Bartholomew, David J. *Mathematical Methods in Social Science.* [25.1]
* ———. *Stochastic Models for Social Processes.* [25.1]
* Bartle, Robert G. *Elements of Real Analysis.* [8.2]
Bartlett, M.S. *The Mathematical Theory of the Dynamics of Biological Populations.* [23.5]
** ———. *Stochastic Population Models in Ecology and Epidemiology.* [23.5]
Bartsch, Hans-Jochen. *Handbook of Mathematical Formulas.* [2.2]
* Barwise, Jon. *The Language of First-Order Logic.* [9.3]
* ———. *The Liar: An Essay on Truth and Circularity.* [9.2]
*** ———. *Handbook of Mathematical Logic.* [9.3]
Bass, Hyman. *Algebraic K-Theory.* [13.13]
Baston, R.J. *Twistors in Mathematics and Physics.* [24.13]
Batastini, Peggy H. *Pharmacological Calculations for Nurses: A Worktext.* [16.3]
* Batchelor, Bruce G. *Practical Approach to Pattern Classification.* [22.11]
*** Batchelor, George K. *Introduction to Fluid Dynamics.* [24.9]
* Bates, Douglas M. *Nonlinear Regression Analysis and Its Applications.* [21.9]
* Bathe, Klaus-Jürgen. *Numerical Methods in Finite Element Analysis.* [18.3]
Batschelet, Edward. *Circular Statistics in Biology.* [23.2]
———. *Introduction to Mathematics for Life Sciences.* [23.1]
Batten, Lynn Margaret. *Combinatorics of Finite Geometries.* [14.9]
Batten, R.W. *Mortality Table Construction.* [17.8]
Bauer, Heinz. *Probability Theory and Elements of Measure Theory.* [20.6]
Baum, Joan. *The Calculating Passion of Ada Byron.* [3.2]
** Baumgart, John K. *Historical Topics for the Mathematics Classroom.* [5.5]
Baumgartner, James E. *Axiomatic Set Theory.* [9.5]
Baumslag, Gilbert. *Reviews on Infinite Groups, 1940–1970.* [2.6]
Bavel, Zamir. *Introduction to the Theory of Automata.* [22.15]
* Baylis, John. *Alice in Numberland: A Students' Guide to the Enjoyment of Higher Mathematics.* [1.4]
Beals, R. *Advanced Mathematical Analysis.* [8.3]
Beard, R.E. *Risk Theory.* [17.7]
* Beasley, John D. *The Ins and Outs of Peg Solitaire.* [4.2]
Beatty, John C. *An Introduction to Splines for Use in Computer Graphics and Geometric Modeling.* [22.18]
Beatty, William E. *Mathematical Relationships in Business and Economics.* [17.2]
Beauchamp, Murray A. *Elements of Mathematical Sociology.* [25.5]
Beauregard, Raymond A. *Linear Algebra.* [12.1]
Beauzamy, Bernard. *Introduction to Banach Spaces and Their Geometry.* [8.8]
* Beck, Anatole. *Excursions Into Mathematics.* [1.4]
Beck, Leland L. *System Software.* [22.10]
* Beck, Robert E. *Elementary Linear Programming with Applications.* [19.5]
** Beckenbach, Edwin F. *An Introduction to Inequalities.* [8.11]
———. *Inequalities.* [8.11]
Becker, Richard A. *The New S Language: A Programming Environment for Data Analysis and Graphics.* [21.8]
Beckmann, Petr. *A History of π(pi).* [3.7]
* Bedini, Silvio. *The Life of Benjamin Banneker.* [3.2]
Beekman, John A. *Actuarial Projections for the Old Age, Survivor, and Disability Income Program of Social Security in the United States of America.* [17.8]
*** ———. *Demography Through Problems.* [23.5]
Beeson, Michael J. *Foundations of Constructive Mathematics: Metamathematical Studies.* [9.8]
Behnke, H. *Fundamentals of Mathematics.* [1.1]
** Behr, Merlyn. *Number Concepts and Operations in the Middle Grades.* [5.12]
* Beiler, Albert. *Recreations in the Theory of Numbers.* [4.1]
Beineke, Lowell W. *Selected Topics in Graph Theory.* [10.5]
Beker, Henry. *Cipher Systems: The Protection of Communications.* [10.6]
* Belhoste, Bruno. *Augustin-Louis Cauchy: A Biography.* [3.2]
Bell, A.W. *A Review of Research in Mathematical Education: Research on Learning and Teaching.* [5.12]
* Bell, C. Gordon. *Computer Structures: Principles and Examples.* [22.17]
Bell, Eric T. *Development of Mathematics.* [3.1]
———. *Mathematics: Queen and Servant of Science.* [1.2]
** ———. *Men of Mathematics.* [3.2]
Bell, J.L. *A Course in Mathematical Logic.* [9.3]
———. *Boolean-Valued Models and Independence Proofs in Set Theory.* [9.5]
Bell, Robbie. *Board Games Round the World: A Resource Book for Mathematical Investigations.* [4.2]
** Bellman, Richard E. *An Introduction to Inequalities.* [8.11]
———. *Analytic Number Theory: An Introduction.* [11.6]
———. *Inequalities.* [8.11]
———. *Stability Theory of Differential Equations.* [7.2]
Belsley, David A. *Regression Diagnostics: Identifying Influential Data and Sources of Collinearity.* [21.9]
* Beltrami, Edward J. *Mathematics for Dynamic Modeling.* [19.2]
———. *Models for Public Systems Analysis.* [25.3]
* Benacerraf, Paul. *Philosophy of Mathematics: Selected Readings.* [9.4]

Blahut, Richard E. *Theory and Practice of Error Control Codes.* [10.6]

Blake, Ian F. *An Introduction to Algebraic and Combinatorial Coding Theory.* [10.6]

Blalock, Hubert M., Jr. *Social Statistics.* [21.3]

* Bleicher, Michael N. *Excursions Into Mathematics.* [1.4]

Bleistein, Norman. *Mathematical Methods for Wave Phenomena.* [7.4]

Bleloch, Andrew L. *How to Model It: Problem Solving for the Computer Age.* [19.2]

Bliss, Elizabeth. *Data Processing Mathematics.* [16.4]

Bloch, R.F. *Mathematical Methods in Medicine.* [23.7]

Bloom, David M. *Linear Algebra and Geometry.* [12.1]

Bluman, George W. *Problem Book for First Year Calculus.* [6.5]

* Blumenthal, Leonard M. *Modern View of Geometry.* [14.7]

Blyth, T.S. *Categories.* [13.10]

Boas, Mary L. *Mathematical Methods in the Physical Sciences.* [24.2]

** Boas, Ralph P., Jr. *A.J. Lohwater's Russian-English Dictionary of the Mathematical Sciences.* [2.2]

*** ———. *A Primer of Real Functions.* [8.3]

* ———. *Invitation to Complex Analysis.* [8.6]

* Bochner, Salomon. *The Role of Mathematics in the Rise of Science.* [1.3]

* Boehm, George A.W. *The Mathematical Sciences: A Collection of Essays.* [1.1]

Boes, D.C. *Introduction to the Theory of Statistics.* [21.4]

* Bogart, Kenneth P. *Discrete Mathematics.* [10.1]

———. *Introductory Combinatorics.* [10.3]

* Boisen, M.B., Jr. *Mathematical Crystallography.* [24.13]

* ———. *Calculus with Applications.* [17.4]

* Boisselle, A.H. *Business Mathematics Today.* [17.1]

Bolc, L. *Expert System Applications.* [22.11]

Bold, Benjamin. *Famous Problems of Geometry and How to Solve Them.* [14.1]

* Bollobás, Béla. *Littlewood's Miscellany.* [1.6]

———. *Combinatorics: Set Systems, Hypergraphs, Families of Vectors, and Combinatorial Probability.* [10.4]

* ———. *Graph Theory: An Introductory Course.* [10.5]

———. *Linear Analysis.* [8.8]

———. *Random Graphs.* [10.5]

* Bolter, J. David. *Turing's Man: Western Culture in the Computer Age.* [22.2]

* Boltyanskii, Vladimir G. *Results and Problems in Combinatorial Geometry.* [14.9]

* ———. *The Decomposition of Figures into Smaller Parts.* [14.9]

* ———. *Convex Figures.* [14.9]

* ———. *Equivalent and Equidecomposable Figures.* [14.9]

** ———. *Hilbert's Third Problem.* [14.9]

———. *Mathematical Methods of Optimal Control.* [19.10]

*** Bondy, J. Adrian. *Graph Theory with Applications.* [10.5]

** Bonnesen, T. *Theory of Convex Bodies.* [14.9]

Bonnet, Robert. *Handbook of Boolean Algebras.* [13.13]

* Bonola, Roberto. *Non-Euclidean Geometry: A Critical and Historical Study of Its Development.* [3.10]

* Booch, Grady. *Software Engineering with Ada.* [22.14]

* Boole, George. *An Investigation of the Laws of Thought.* [9.2]

———. *Treatise on the Calculus of Finite Differences.* [3.4]

* Boolos, George S. *Computability and Logic.* [9.3]

———. *The Unprovability of Consistency: An Essay in Modal Logic.* [9.3]

** Boot, John C.G. *Common Globe or Global Commons: Population Regulation and Income Distribution.* [25.2]

* Booth, Taylor L. *Digital Networks and Computer Systems.* [22.17]

* Boothby, William M. *An Introduction to Differentiable Manifolds and Riemannian Geometry.* [14.10]

Boots, Barry. *Models of Spatial Processes: An Approach to the Study of Point, Line, and Area Patterns.* [19.3]

Borel, Émile. *Elements of the Theory of Probability.* [20.1]

*** Borevich, Z.I. *Number Theory.* [11.5]

* Borisenko, A.I. *Vector and Tensor Analysis with Applications.* [24.11]

Borovkov, A.A. *Asymptotic Methods in Queueing Theory.* [19.8]

———. *Stochastic Processes in Queueing Theory.* [19.8]

Borowski, E.J. *Collins Reference Dictionary of Mathematics.* [2.1]

Borrie, M.S. *Modelling with Differential Equations.* [7.1]

Borwein, Jonathan M. *A Dictionary of Real Numbers.* [2.2]

* ———. *Pi and the AGM: A Study in Analytic Number Theory and Computational Complexity.* [11.6]

———. *Collins Reference Dictionary of Mathematics.* [2.1]

Borwein, Peter B. *A Dictionary of Real Numbers.* [2.2]

* ———. *Pi and the AGM: A Study in Analytic Number Theory and Computational Complexity.* [11.6]

Bose, R.C. *Introduction to Combinatorial Theory.* [10.3]

* Bottazzini, Umberto. *The Higher Calculus: A History of Real and Complex Analysis from Euler to Weierstrass.* [3.9]

* Bourbaki, Nicolas. *Elements of Mathematics: Algebra.* [13.4]

* ———. *Elements of Mathematics: Commutative Algebra.* [13.8]

———. *Elements of Mathematics: General Topology.* [15.1]

Brown, William G. *Reviews in Graph Theory, 1940–1978.* [2.6]
Brualdi, Richard A. *Combinatorial Matrix Theory.* [10.4]
** ———. *Introductory Combinatorics.* [10.3]
Bruce, J.W. *Curves and Singularities: A Geometrical Introduction to Singularity Theory.* [14.11]
———. *Microcomputers and Mathematics.* [22.1]
* Bruni, James V. *Experiencing Geometry.* [14.3]
Bryant, Victor. *The Sunday Times Book of Brain Teasers.* [4.3]
Bucciarelli, Louis L. *Sophie Germain: An Essay in the History of the Theory of Elasticity.* [3.2]
Buchanan, J.T. *Discrete and Dynamic Decision Analysis.* [19.5]
** Buck, R. Creighton. *Advanced Calculus.* [6.4]
Buckley, Fred. *Distance in Graphs.* [10.5]
** Budden, F.J. *The Fascination of Groups.* [13.5]
** Budnick, Frank S. *Applied Mathematics for Business, Economics, and the Social Sciences.* [17.2]
Buell, Duncan A. *Binary Quadratic Forms: Classical Theory and Modern Computations.* [11.5]
Buerger, David J. *LaTeX for Scientists and Engineers.* [22.13]
** Bühler, Walter K. *Gauss: A Biographical Study.* [3.2]
Buhlman, Hans. *Mathematical Methods in Risk Theory.* [17.7]
** Bulirsch, R. *Introduction to Numerical Analysis.* [18.1]
Bulmer, M.G. *The Mathematical Theory of Quantitative Genetics.* [23.4]
Bumcrot, Robert J. *Introduction to Discrete Mathematics.* [10.1]
Bunch, Bryan H. *Mathematical Fallacies and Paradoxes.* [1.2]
* ———. *Reality's Mirrors: Exploring the Mathematics of Symmetry.* [14.5]
Buncher, C. Ralph. *Statistics in the Pharmaceutical Industry.* [21.16]
* Bundy, Alan. *The Computer Modelling of Mathematical Reasoning.* [22.11]
Burckel, Robert B. *An Introduction to Classical Complex Analysis.* [8.6]
Burde, Gerhard. *Knots.* [15.2]
*** Burden, Richard L. *Numerical Analysis.* [18.1]
Burger, Dionys. *Sphereland.* [14.1]
Burghes, David N. *Modelling with Differential Equations.* [7.1]
———. *Applying Mathematics: A Course in Mathematical Modelling.* [19.2]
* Burgmeier, J.W. *Calculus with Applications.* [17.4]
** Burington, Richard S. *Handbook of Probability and Statistics with Tables.* [2.2]
*** Burke, Charles J. *Applied Mathematics for Business, Economics, Life Sciences, and Social Sciences.* [17.2]
* Burke, William L. *Applied Differential Geometry.* [14.10]
Burkill, John C. *A First Course in Mathematical Analysis.* [8.2]
———. *The Lebesgue Integral.* [8.3]
Burkinshaw, Owen. *Principles of Real Analysis.* [8.2]
Burks, Alice R. *The First Electronic Computer: The Atanasoff Story.* [3.2]
Burks, Arthur W. *The First Electronic Computer: The Atanasoff Story.* [3.2]
Burn, R.P. *A Pathway Into Number Theory.* [11.1]
* ———. *Groups: A Path to Geometry.* [13.5]
Burns, Marilyn. *The Book of Think (Or How to Solve a Problem Twice Your Size).* [5.7]
———. *The I Hate Mathematics! Book.* [5.7]
Burnside, William Snow. *The Theory of Equations with an Introduction to the Theory of Binary Algebraic Forms.* [13.1]
Burr, Irving W. *Elementary Statistical Quality Control.* [21.14]
Burris, Stanley. *A Course in Universal Algebra.* [13.12]
* Burstein, Samuel. *Calculus with Applications and Computing.* [6.3]
Burton, David M. *Elementary Number Theory.* [11.1]
* ———. *The History of Mathematics: An Introduction.* [3.1]
Burton, D. *Abstract Algebra.* [13.1]
Burton, L. *Girls Into Maths Can Go.* [5.10]
* Burton, Leone. *Thinking Mathematically.* [5.11]
Burton, T.A. *Modeling and Differential Equations in Biology.* [23.1]
———. *Stability and Periodic Solutions of Ordinary and Functional Differential Equations.* [7.3]
Bustard, David. *Sequential Program Structures.* [22.4]
Butcher, Marjorie V. *Mathematics of Compound Interest.* [17.5]

C

* Cacoullos, T. *Exercises in Probability.* [20.1]
Cadwell, J.H. *Topics in Recreational Mathematics.* [4.1]
Cadzow, James A. *Foundations of Digital Signal Processing and Data Analysis.* [24.12]
** Cajori, Florian. *A History of Mathematical Notations.* [3.1]
* ———. *A History of Mathematics.* [3.1]
———. *The Teaching and History of Mathematics in the United States.* [5.5]
Calder, Nigel. *Einstein's Universe.* [24.7]
California Assessment Program. *A Question of Thinking: A First Look at Students' Performance on Open-ended Questions in Mathematics.* [5.13]
* Calingaert, Peter. *Program Translation Fundamentals: Methods and Issues.* [22.16]
* Calinger, Ronald. *Classics of Mathematics.* [3.3]
———. *Gottfried Wilhelm Leibniz.* [3.2]
* Calter, Paul. *Mathematics for Computer Technology.* [16.4]
** ———. *Technical Mathematics with Calculus.* [16.7]
Calude, Cristian. *Theories of Computational Complexity.* [22.9]
Cameron, P.J. *Graphs, Codes and Designs.* [10.6]
** Cameron, R.G. *Introduction to Mathematical Control Theory.* [19.10]

** Chevalley, Claude. *Theory of Lie Groups.* [8.13]
Chiang, Chin Long. *The Life Table and Its Applications.* [17.8]
Chihara, T.S. *An Introduction to Orthogonal Polynomials.* [8.4]
* Childress, Stephen. *Mechanics of Swimming and Flying.* [23.8]
* Childs, Lindsay. *A Concrete Introduction to Higher Algebra.* [13.1]
Chillingworth, D.R.J. *Differential Topology with a View to Applications.* [15.4]
Chinitz, M. Paul. *The Logic Design of Computers: An Introduction.* [22.17]
** Chinn, William G. *First Concepts of Topology.* [15.1]
* ———. *3.1416 And All That.* [1.6]
Chorin, A.J. *A Mathematical Introduction to Fluid Mechanics.* [24.9]
Chorlton, Frank. *Vector & Tensor Methods.* [24.11]
Chou, Shang-Ching. *Mechanical Geometry Theorem Proving.* [22.11]
Chow, S. *Methods of Bifurcation Theory.* [7.3]
* Chow, Yuan Shih. *Probability Theory: Independence, Interchangeability, Martingales.* [20.3]
———. *Great Expectations: The Theory of Optimal Stopping.* [20.4]
* Christiansen, B. *Perspectives on Mathematics Education.* [5.2]
* Christiansen, F.B. *Theories of Populations in Biological Communities.* [23.4]
Chrystal, George. *Textbook of Algebra.* [3.4]
Chua, Leon O. *Practical Numerical Algorithms for Chaotic Systems.* [18.3]
Chui, C.K. *Linear Systems and Optimal Control.* [19.10]
Chung, Kai Lai. *Introduction to Stochastic Integration.* [20.4]
** ———. *A Course in Probability Theory.* [20.3]
*** ———. *Elementary Probability Theory with Stochastic Processes.* [20.2]
Churchhouse, Robert F. *Numerical Methods.* [18.2]
*** Churchill, Ruel V. *Fourier Series and Boundary Value Problems.* [7.5]
*** Chvátal, Vasek. *Linear Programming.* [19.5]
** Ciarlet, Philippe G. *Numerical Linear Algebra.* [18.4]
———. *The Finite Element Method for Elliptic Problems.* [18.3]
Ciesielski, Zbigniew. *Proceedings of the International Congress of Mathematicians, August 16–24, 1983, Warszawa.* [2.7]
** Cipra, Barry. *Misteaks ... and How to Find Them Before the Teacher Does ....* [6.5]
* Cissell, H. *Mathematics of Finance.* [17.5]
* Cissell, R. *Mathematics of Finance.* [17.5]
* Clark, Colin W. *Bioeconomic Modelling and Fisheries Management.* [25.2]
** ———. *Dynamic Modeling in Behavioral Ecology.* [23.2]
*** ———. *Mathematical Bioeconomics: The Optimal Management of Renewable Resources.* [25.2]
** Clark, Frank J. *Mathematics for Programming Computers.* [16.4]
Clarke, C. *Elementary General Relativity.* [24.7]
* Clason, Robert G. *Readings in the History of Mathematics Education.* [5.5]
Clayson, J. *Visual Modeling with Logo.* [14.8]
Cleary, James P. *The Professional Forecaster: The Forecasting Process Through Data Analysis.* [21.16]
Cleaves, Cheryl. *Introduction to Technical Mathematics.* [16.1]
* ———. *Basic Mathematics for Trades and Technologies.* [16.2]
Clemens, C. Herbert. *A Scrapbook of Complex Curve Theory.* [14.11]
Clemens, Stanley R. *Geometry.* [14.3]
** ———. *Geometry: An Investigative Approach.* [14.3]
Clements, M.A. *School Mathematics: The Challenge to Change.* [5.1]
*** Cleveland, William S. *The Elements of Graphing Data.* [21.6]
** Clocksin, W.F. *Programming in Prolog.* [22.7]
Clooney, Thomas J. *Geometry.* [14.3]
Closs, Michael P. *Native American Mathematics.* [5.4]
Cobb, Paul. *Construction of Arithmetical Meanings and Strategies.* [5.12]
* Cobelli, C. *The Mathematical Modeling of Metabolic and Endocrine Systems.* [23.6]
Cochran, James A. *Applied Mathematics: Principles, Techniques, and Applications.* [24.2]
Cochran, William G. *Sampling Techniques.* [21.13]
*** ———. *Statistical Methods.* [21.4]
* Cockcroft, Wilfred H. *Mathematics Counts.* [5.1]
Cocking, R.R. *Linguistic and Cultural Influences on Learning Mathematics.* [5.4]
Codd, E.F. *Cellular Automata.* [22.15]
** ———. *The Relational Model.* [22.5]
*** Coddington, Earl A. *Theory of Ordinary Differential Equations.* [7.2]
* ———. *An Introduction to Ordinary Differential Equations.* [7.1]
Cohen, A. Toni. *Principles of Computer Science: Concepts, Algorithms, Data Structures, and Applications.* [22.3]
* Cohen, Daniel I.A. *Basic Techniques of Combinatorial Theory.* [10.3]
Cohen, David. *Precalculus.* [6.2]
* Cohen, Joel E. *Food Webs and Niche Space.* [23.2]
Cohen, Jonathan. *An Introduction to the Philosophy of Induction and Probability.* [9.4]
Cohen, J.W. *The Single Server Queue.* [19.8]
* Cohen, Patricia Cline. *A Calculating People: The Spread of Numeracy in Early America.* [5.5]
** Cohen, Paul R. *The Handbook of Artificial Intelligence.* [22.11]
Cohn, Harvey. *A Classical Invitation to Algebraic Numbers and Class Fields.* [11.5]
** Cohn, Paul M. *Algebra.* [13.4]
———. *Free Rings and Their Relations.* [13.6]
* ———. *Universal Algebra.* [13.12]
* Cohn, Victor. *News & Numbers.* [1.6]
*** Cohn-Vossen, S. *Geometry and the Imagination.* [14.1]
* Cohon, Jared L. *Multiobjective Programming and Planning.* [19.6]

** ———. *Projective Geometry.* [14.7]
———. *Regular Polytopes.* [14.5]
* ———. *The Real Projective Plane.* [14.7]
Coxford, Arthur F. *Advanced Mathematics: A Preparation for Calculus.* [6.2]
*** ———. *The Ideas of Algebra, K–12: 1988 Yearbook.* [5.8]
Cozzens, Margaret B. *Mathematics and Its Applications to Management, Life, and Social Sciences With Finite and Discrete Mathematics.* [10.2]
Craig, Allen T. *Introduction to Mathematical Statistics.* [21.4]
Cramér, Harald. *Stationary and Related Stochastic Processes.* [20.4]
———. *Random Variables and Probability Distributions.* [20.3]
———. *The Elements of Probability Theory.* [20.2]
Crank, John. *Free and Moving Boundary Problems.* [7.5]
** Croft, Hallard T. *Unsolved Problems in Geometry.* [14.1]
* Croft, Tony. *Modelling with Projectiles.* [24.13]
Cronin, T.M. *Mathematical Formulae for Engineering and Science Students.* [2.2]
Cronin-Scanlon, Jane. *Advanced Calculus: A Start in Analysis.* [6.4]
Cross, Mark. *Learning the Art of Mathematical Modelling.* [19.2]
** Crossley, J.N. *What is Mathematical Logic?* [9.3]
———. *The Emergence of Number.* [3.7]
** Crow, James R. *An Introduction to Population Genetics Theory.* [23.4]
* Crowe, Donald W. *Excursions Into Mathematics.* [1.4]
** ———. *Symmetries of Culture: Handbook of Plane Pattern Analysis.* [25.1]
Crowe, Michael J. *A History of Vector Analysis: The Evolution of the Idea of a Vectorial System.* [3.9]
* Crowell, Richard H. *Introduction to Knot Theory.* [15.2]
———. *Calculus of Vector Functions.* [6.4]
Crowley, Michael F. *Women and Minorities in Science and Engineering.* [5.10]
* Crown, J. Conrad. *Finite Mathematics.* [10.2]
** ———. *Mathematics and Calculus with Applications.* [17.4]
Crump, Thomas. *The Anthropology of Numbers.* [3.7]
Cryer, Jonathan D. *Time Series Analysis.* [21.15]
* Cullen, Michael R. *Mathematics for the Biosciences.* [23.1]
** Cundy, M.H. *Mathematical Models.* [14.5]
* Cunningham, Steve. *Visualization in Teaching and Learning Mathematics.* [5.9]
———. *Programming the User Interface.* [22.14]
Curry, Haskell B. *Foundations of Mathematical Logic.* [9.3]
Curtis, A.B. *Mathematics of Accounting.* [17.5]
* Curtis, Charles W. *Methods of Representation Theory with Applications to Finite Groups and Orders.* [13.5]
* ———. *Representation Theory of Finite Groups and Associative Algebras.* [13.5]
* ———. *Linear Algebra: An Introductory Approach.* [12.1]
Cutland, Nigel. *Computability: An Introduction to Recursive Function Theory.* [9.6]

D

** D'Abrera, H.J.M. *Nonparametrics: Statistical Methods Based on Ranks.* [21.11]
* Daellenbach, Hans G. *Introduction to Operations Research Techniques.* [19.1]
Dahl, O.J. *Structured Programming.* [22.6]
Dahlquist, Germund. *Numerical Methods.* [18.2]
* Daintith, John. *The Penguin Dictionary of Mathematics.* [2.1]
Dale, Nell. *Pascal Plus Data Structures.* [22.4]
Dales, H.G. *An Introduction to Independence for Analysts.* [9.5]
Dalton, LeRoy C. *Topics for Mathematics Clubs.* [5.8]
Damiano, David B. *A Course in Linear Algebra.* [12.1]
* Daniel, James W. *Applied Linear Algebra.* [12.1]
Daniels, Jeanne M. *Clinical Calculations: A Unified Approach.* [16.3]
* Dantzig, George B. *Studies in Optimization.* [19.1]
*** ———. *Linear Programming and Extensions.* [19.5]
Dantzig, Tobias. *Number, The Language of Science.* [1.2]
Daston, Lorraine. *Classical Probability in the Enlightenment.* [3.11]
*** Date, C.J. *An Introduction to Database Systems.* [22.5]
** ———. *Relational Database Writings.* [22.5]
Dauben, Joseph W. *Mathematical Perspectives: Essays on Mathematics and Its Historical Development.* [3.6]
* ———. *Georg Cantor: His Mathematics and Philosophy of the Infinite.* [3.2]
** ———. *The History of Mathematics from Antiquity to the Present: A Selective Bibliography.* [2.4]
** Davenport, Harold. *Multiplicative Number Theory.* [11.6]
*** ———. *The Higher Arithmetic: An Introduction to the Theory of Numbers.* [11.1]
* Davenport, J.H. *Computer Algebra: Systems and Algorithms for Algebraic Computation.* [22.8]
Davey, B.A. *Introduction to Lattices and Order.* [10.7]
* David, F.N. *Games, Gods, and Gambling.* [20.1]
David, H.A. *Order Statistics.* [21.5]
Davidson, Judy K. *Pharmacological Calculations for Nurses: A Worktext.* [16.3]
* Davidson, Neil. *Cooperative Learning in Mathematics: A Handbook for Teachers.* [5.6]
Davie, A.J.T. *Recursive Descent Compiling.* [22.16]
Davies, B. *Integral Transforms and their Applications.* [8.4]
Davies, Glyn A.O. *Mathematical Methods in Engineering.* [24.2]

———. *Sets, Functions, and Logic: Basic Concepts of University Mathematics.* [6.2]
———. *The Axiom of Constructibility: A Guide for the Mathematician.* [9.5]
* Devore, Jay L. *Statistics: The Exploration and Analysis of Data.* [21.2]
———. *Probability and Statistics for Engineering and the Sciences.* [21.4]
* DeVore, Russell. *Practical Problems in Mathematics for Heating and Cooling Technicians.* [16.2]
Devroye, L. *Non-Uniform Random Variate Generation.* [19.9]
*** Dewdney, A.K. *The Armchair Universe: An Exploration of Computer Worlds.* [22.1]
** ———. *The Planiverse.* [1.5]
** ———. *The Turing Omnibus: 61 Excursions in Computer Science.* [22.1]
Dewhurst, Stephen C. *Programming in C++.* [22.7]
DeWitt, Bryce S. *The Many-Worlds Interpretation of Quantum Mechanics.* [24.5]
** Diamond, H.G. *The Theory of Algebraic Numbers.* [11.5]
* Diamond, J.M. *Community Ecology.* [23.2]
** Dick, Auguste. *Emmy Noether, 1882–1935.* [3.2]
Dick, Elie M. *Current Information Sources in Mathematics: An Annotated Guide to Books and Periodicals, 1960–1972.* [2.4]
Dickson, Leonard E. *Algebraic Numbers.* [11.5]
*** ———. *History of the Theory of Numbers.* [3.7]
———. *Introduction to the Theory of Numbers.* [11.1]
———. *Modern Elementary Theory of Numbers.* [11.3]
———. *Studies in the Theory of Numbers.* [11.5]
Diekmann, O. *The Dynamics of Physiologically Structured Populations.* [23.5]
* Dietrich, Frank H. *Statistics.* [21.2]
* Dieudonné, Jean. *A Panorama of Pure Mathematics as Seen by N. Bourbaki.* [1.3]
———. *A History of Algebraic and Differential Topology, 1900–1960.* [3.10]
———. *Abrégé d'histoire des Mathématiques, 1700–1900.* [3.1]
———. *History of Algebraic Geometry.* [3.10]
———. *History of Functional Analysis.* [3.9]
———. *Linear Algebra and Geometry.* [12.2]
———. *Special Functions and Linear Representations of Lie Groups.* [8.13]
———. *Treatise on Analysis.* [8.8]
* Diggle, Peter J. *Time Series: A Biostatistical Approach.* [21.15]
** Dijksterhuis, E.J. *Archimedes.* [3.4]
* Dijkstra, Edsger W. *A Method of Programming.* [22.8]
———. *Structured Programming.* [22.6]
* ———. *A Discipline of Programming.* [22.6]
*** DiPrima, Richard C. *Elementary Differential Equations and Boundary Value Problems.* [7.1]
** diSessa, Andrea A. *Turtle Geometry: The Computer as a Medium for Exploring Mathematics.* [14.8]
Dixon, J.D. *Problems in Group Theory.* [13.5]
** do Carmo, Manfredo P. *Differential Geometry of Curves and Surfaces.* [14.10]
* Dobbins, Robert R. *If And Only If in Analysis.* [6.5]
* Dobbs, David E. *A Modern Course on the Theory of Equations.* [13.2]
Dobson, Annette J. *An Introduction to Generalized Linear Models.* [21.9]
* Dodson, C.T.J. *Tensor Geometry: The Geometric Viewpoint and its Uses.* [14.10]
Doerr, Alan. *Applied Discrete Structures for Computer Science.* [10.1]
Doig, A.G. *Bibliography of Statistical Literature.* [2.4]
* Doksum, Kjell A. *Mathematical Statistics: Basic Ideas and Selected Topics.* [21.4]
Domoryad, A.P. *Mathematical Games and Pastimes.* [4.2]
Doob, J.L. *Classical Potential Theory and Its Probabilistic Counterpart.* [20.8]
*** ———. *Stochastic Processes.* [20.4]
Doob, Michael. *A Gentle Introduction to TEX.* [22.13]
Doran, J.E. *Mathematics and Computers in Archaeology.* [25.1]
Dorling, A.R. *Use of Mathematical Literature.* [2.4]
Dorn, William S. *Numerical Methods with FORTRAN IV Case Studies.* [18.1]
———. *Computer Applications for Calculus.* [6.3]
* Dornhoff, Larry L. *Applied Modern Algebra.* [13.3]
Dörrie, Heinrich. *100 Great Problems of Elementary Mathematics.* [1.3]
Dorwart, Harold L. *Geometry of Incidence.* [14.7]
Dossey, John A. *Discrete Mathematics.* [10.1]
** ———. *The Mathematics Report Card: Are We Measuring Up?* [5.13]
———. *Contemporary Mathematics.* [1.4]
** Douglas, Ronald G. *Toward A Lean and Lively Calculus.* [5.9]
Downs, Floyd L. *Geometry.* [14.3]
* Doyle, Peter G. *Random Walks and Electric Networks.* [20.4]
** Draper, Norman R. *Applied Regression Analysis.* [21.9]
Dresher, Melvin. *The Mathematics of Games of Strategy: Theory and Applications.* [19.4]
** Drew, Donald A. *Differential Equation Models.* [19.2]
Dreyfus, Hubert L. *What Computers Can't Do.* [22.2]
Dreyfus, Stuart E. *The Art and Theory of Dynamic Programming.* [19.3]
Driscoll, Mark J. *Teaching Mathematics: Strategies that Work.* [5.6]
———. *Research Within Reach: Elementary School Mathematics.* [5.12]
Driver, Rodney D. *Ordinary and Delay Differential Equations.* [7.2]
** Dù Shírán. *Chinese Mathematics: A Concise History.* [3.5]
Dubins, Lester E. *Inequalities for Stochastic Processes (How to Gamble If You Must).* [20.4]
Duce, David A. *Distributed Computing.* [22.10]
*** Dudeney, Henry E. *536 Puzzles and Curious Problems.* [4.3]
** ———. *Amusements in Mathematics.* [4.3]
———. *The Canterbury Puzzles.* [4.3]

* ———. *An Introduction to Mathematical Taxonomy.* [23.2]
*** Eves, Howard W. *A Survey of Geometry.* [14.2]
*** ———. *An Introduction to the History of Mathematics with Cultural Connections.* [3.1]
* ———. *Foundations and Fundamental Concepts of Mathematics.* [9.1]
* ———. *Great Moments in Mathematics (After 1650).* [3.6]
* ———. *Great Moments in Mathematics (Before 1650).* [3.5]
** ———. *Mathematical Circles.* [1.5]
** Ewen, Dale. *Elementary Technical Mathematics.* [16.1]
* ———. *Mathematics for Technical Education.* [16.7]
* ———. *Technical Calculus.* [16.7]
*** Ewens, W.J. *Mathematical Population Genetics.* [23.4]
* Ewing, John. *Puzzle It Out: Cubes, Groups and Puzzles.* [4.2]

F

Faber, Richard L. *Differential Geometry and Relativity Theory: An Introduction.* [14.10]
* ———. *Foundations of Euclidean and Non-Euclidean Geometry.* [14.4]
* Fadiman, Clifton. *Fantasia Mathematica.* [1.5]
* ———. *The Mathematical Magpie.* [1.5]
*** Faires, J. Douglas. *Numerical Analysis.* [18.1]
** Falconer, Kenneth J. *Unsolved Problems in Geometry.* [14.1]
———. *The Geometry of Fractal Sets.* [8.5]
* Fararo, Thomas J. *Mathematical Sociology: An Introduction to Fundamentals.* [25.5]
Farlow, Stanley J. *Applied Mathematics for Business, Economics, and the Social Sciences.* [17.2]
* ———. *Calculus and Its Applications.* [17.4]
———. *Finite Mathematics and Its Applications.* [10.2]
———. *Partial Differential Equations for Scientists and Engineers.* [7.4]
Farnum, Nicholas R. *Quantitative Forecasting Methods.* [21.5]
*** Fauvel, John. *The History of Mathematics: A Reader.* [3.1]
* ———. *Let Newton Be! A New Perspective on His Life and Works.* [3.2]
Federico, P.J. *Descartes on Polyhedra: A Study of the De Solidorum Elementis.* [3.4]
Feferman, Solomon. *The Number Systems: Foundations of Algebra and Analysis.* [8.1]
Feigelstock, S. *Additive Groups of Rings.* [13.5]
** Feigenbaum, Edward A. *The Handbook of Artificial Intelligence.* [22.11]
* Feijen, W.H.J. *A Method of Programming.* [22.8]
Feinerman, Robert P. *Polynomial Approximation.* [18.5]
** Feldman, M.W. *Mathematical Evolutionary Theory.* [23.4]
———. *Cultural Transmission and Evolution: A Quantitative Approach.* [23.4]
Feldman, Norman. *Introduction to Advanced Mathematics.* [8.1]
Felker, C.A. *Shop Mathematics.* [16.2]
*** Feller, William. *An Introduction to Probability Theory and Its Applications.* [20.3]
* Fenchel, T.M. *Theories of Populations in Biological Communities.* [23.4]
** Fenchel, W. *Theory of Convex Bodies.* [14.9]
* Fendel, Daniel. *Foundations of Higher Mathematics: Exploration and Proof.* [8.1]
Fennell, Francis. *Ideas from the Arithmetic Teacher, Grades 4–6: Intermediate School.* [5.7]
** Fennema, Elisabeth. *Mathematics and Gender.* [5.10]
Feroe, John. *Single Variable Calculus with Discrete Mathematics.* [6.3]
* Fetisov, A.I. *Proof in Geometry.* [14.3]
Feuer, A. *Comparing and Assessing Programming Languages: Ada, C, Pascal.* [22.6]
Fey, James T. *Computing and Mathematics: The Impact on Secondary School Curricula.* [5.14]
* Fienberg, Stephen E. *Beginning Statistics with Data Analysis.* [21.6]
* ———. *Statistics and the Law.* [21.1]
* ———. *The Analysis of Cross-Classified Categorical Data.* [21.6]
Findley, A.M. *The Geometry of Genetics.* [23.4]
Findley, G.L. *The Geometry of Genetics.* [23.4]
* Fine, Nathan J. *Basic Hypergeometric Series and Applications.* [8.15]
Fine, Terrence L. *Theories of Probability: An Examination of Foundations.* [20.5]
Finkbeiner, Daniel T. *A Primer of Discrete Mathematics.* [10.1]
* Finkelstein, L. *The Mathematical Modeling of Metabolic and Endocrine Systems.* [23.6]
* Finkelstein, Michael O. *Statistics for Lawyers.* [21.1]
* Finney, Ross L. *Calculus.* [6.3]
*** ———. *Calculus and Analytic Geometry.* [6.3]
** Firby, P.A. *Surface Topology.* [15.2]
Firschein, Oscar. *Intelligence: The Eye, the Brain, and the Computer.* [22.11]
Fischbein, E. *The Intuitive Sources of Probabilistic Thinking in Children.* [5.3]
*** Fischer, Charles N. *Crafting a Compiler.* [22.16]
Fischer, Emanuel. *Intermediate Real Analysis.* [8.3]
* Fischer, Gerd. *Mathematical Models from the Collections of Universities and Museums.* [14.1]
Fischler, Martin A. *Intelligence: The Eye, the Brain, and the Computer.* [22.11]
Fishback, William T. *Projective and Euclidean Geometry.* [14.7]
* Fishburn, Peter C. *Approval Voting.* [25.3]
———. *The Theory of Social Choice.* [25.3]
** Fisher, Paul. *Collegiate Business Mathematics.* [17.1]
** Fisher, Ronald A. *Statistical Methods, Experimental Design, and Scientific Inference.* [21.12]
Fisher, Stephen D. *Complex Variables.* [8.7]
** Fishman, G.S. *Principles of Discrete Event Simulation.* [19.9]

G

** Gaal, Lisl. *Classical Galois Theory with Examples.* [13.7]

* Gaffney, Matthew P. *Annotated Bibliography of Expository Writing in the Mathematical Sciences.* [2.4]

* Gale, David. *The Theory of Linear Economic Models.* [25.2]

* Gale, J.S. *Theoretical Population Genetics.* [23.4]

Galilei, Galileo. *Dialogues Concerning Two New Sciences.* [3.4]

** Gallian, Joseph A. *Contemporary Abstract Algebra.* [13.1]

Gallin, Daniel. *Applied Mathematics for the Management, Life, and Social Sciences.* [17.2]

———. *Finite Mathematics.* [10.2]

* Gallivan, K.A. *Parallel Algorithms for Matrix Computations.* [22.8]

* Galvin, Peter B. *Operating System Concepts.* [22.10]

Gamelin, Theodore W. *Introduction to Topology.* [15.1]

———. *Uniform Algebras.* [8.8]

* Gani, J. *The Craft of Probabilistic Modelling: A Collection of Personal Accounts.* [20.1]

* ———. *The Making of Statisticians.* [3.11]

* Gans, David. *An Introduction to Non-Euclidean Geometry.* [14.4]

* ———. *Transformations and Geometries.* [14.6]

Gantmacher, Felix. *Matrix Theory.* [12.3]

** Garabedian, Paul R. *Partial Differential Equations.* [7.4]

Garden, Robert A. *The IEA Study of Mathematics II: Contexts and Outcomes of School Mathematics.* [5.13]

** Gardiner, A. *Discovering Mathematics: The Art of Investigation.* [1.4]

* ———. *Infinite Processes: Background to Analysis.* [8.1]

* ———. *Mathematical Puzzling.* [4.6]

** Gardiner, C.F. *Surface Topology.* [15.2]

Gårding, Lars. *Encounter with Mathematics.* [1.2]

*** Gardner, Martin. *The Mathematical Puzzles of Sam Loyd.* [4.3]

———. *Aha! Gotcha.* [4.5]

———. *Aha! Insight.* [4.5]

* ———. *Hexaflexagons and Other Mathematical Diversions.* [4.5]

* ———. *Knotted Doughnuts and Other Mathematical Entertainments.* [4.5]

* ———. *Martin Gardner's New Mathematical Diversions from Scientific American.* [4.5]

* ———. *Martin Gardner's Sixth Book of Mathematical Diversions from Scientific American.* [4.5]

* ———. *Mathematical Carnival.* [4.5]

* ———. *Mathematical Circus: More Games, Puzzles, Paradoxes, and Other Mathematical Entertainments from Scientific American.* [4.5]

* ———. *Mathematical Magic Show.* [4.5]

———. *Mathematics, Magic and Mystery.* [4.5]

* ———. *Penrose Tiles to Trapdoor Ciphers.* [4.5]

* ———. *Riddles of the Sphinx And Other Mathematical Puzzle Tales.* [4.5]

———. *Science Fiction Puzzle Tales.* [4.5]

* ———. *The Magic Numbers of Dr. Matrix.* [4.5]

* ———. *The Scientific American Book of Mathematical Puzzles and Diversions.* [4.5]

* ———. *The Second Scientific American Book of Mathematical Puzzles and Diversions.* [4.5]

* ———. *Time Travel and Other Mathematical Bewilderments.* [4.5]

* ———. *Wheels, Life, and Other Mathematical Amusements.* [4.5]

*** Garey, Michael R. *Computers and Intractability: A Guide to the Theory of NP-Completeness.* [22.9]

Garfinkel, Robert S. *Integer Programming.* [19.7]

Garling, D.J.H. *A Course in Galois Theory.* [13.7]

* Garner, Cyril W.L. *n-gons.* [13.13]

Garner, Lynn E. *An Outline of Projective Geometry.* [14.7]

* Garrad, Crawford G. *Practical Problems in Mathematics for Electricians.* [16.2]

* Garrett, Leonard J. *Data Structures, Algorithms, and Program Style Using C.* [22.4]

Gasper, G. *Basic Hypergeometric Series.* [8.15]

Gass, Saul I. *Operations Research: Mathematics and Models.* [19.2]

* ———. *Decision Making, Models, and Algorithms: A First Course.* [19.1]

** ———. *Linear Programming: Methods and Applications.* [19.5]

* Gause, G.F. *The Struggle for Existence.* [23.2]

* Gauss, Carl Friedrich. *Disquisitiones Arithmeticae.* [3.4]

Gawronski, J.D. *Changing School Mathematics: A Responsive Process.* [5.1]

** Gazis, Denos C. *Traffic Science.* [19.3]

Geddes, Dorothy. *The Van Hiele Model of Thinking in Geometry Among Adolescents.* [5.3]

Gehani, Narain. *Comparing and Assessing Programming Languages: Ada, C, Pascal.* [22.6]

* ———. *C: An Advanced Introduction.* [22.7]

Geis, I. *How to Lie with Statistics.* [21.1]

———. *How to Take a Chance.* [20.1]

*** Gelbaum, Bernard R. *Theorems and Counterexamples in Mathematics.* [8.3]

Gelenbe, E. *Introduction to Queueing Networks.* [19.8]

* Gel'fand, Israel M. *Generalized Functions.* [8.8]

* ———. *Commutative Normed Rings.* [8.8]

* ———. *Lectures on Linear Algebra.* [12.2]

* Gelfond, A.O. *Transcendental and Algebraic Numbers.* [11.5]

* Gellert, Walter. *The VNR Concise Encyclopedia of Mathematics.* [2.3]

Gemignani, Michael C. *Elementary Topology.* [15.1]

Gentle, James E. *Statistical Computing.* [21.8]

* George, John A. *Introduction to Operations Research Techniques.* [19.1]

* Gerald, Curtis F. *Applied Numerical Analysis.* [18.1]

———. *Computer Graphics: The Principles Behind the Art and Science.* [22.18]

Geramita, Anthony V. *Introduction to Homological Methods in Commutative Rings.* [13.9]

Gerber, Hans U. *An Introduction to Mathematical Risk Theory.* [17.7]

Golenberg, E.M. *Lectures in Theoretical Population Biology.* [23.5]
** Golomb, Solomon W. *Polyominoes.* [4.6]
Golos, Ellery B. *Patterns in Mathematics.* [1.4]
*** Golub, Gene H. *Matrix Computations.* [18.4]
———. *Reviews in Numerical Analysis, 1980–1986.* [2.6]
** ———. *Studies in Numerical Analysis.* [18.2]
* Gonshor, Harry. *An Introduction to the Theory of Surreal Numbers.* [9.9]
Good, I.J. *Good Thinking: The Foundations of Probability and Its Applications.* [20.5]
Goodbody, A.M. *Cartesian Tensors: With Applications to Mechanics, Fluid Mechanics and Elasticity.* [12.4]
* Goodearl, K.R. *An Introduction to Noncommutative Noetherian Rings.* [13.6]
* Goodier, J.N. *Theory of Elasticity.* [24.4]
Goodson, C.E. *Technical Mathematics with Calculus.* [16.7]
Goodstein, R.L. *Development of Mathematical Logic.* [9.3]
Goos, Gerhard. *Compiler Construction.* [22.16]
Goppa, V.D. *Geometry and Codes.* [10.6]
Gõranzon, Bo. *Knowledge, Skill, and Artificial Intelligence.* [22.11]
Gordon, A.D. *Classification: Methods for the Exploratory Analysis of Multivariate Data.* [21.10]
Gordon, V.O. *A Course in Descriptive Geometry.* [14.12]
Gorenstein, Daniel. *Reviews on Finite Groups, 1940–1970.* [2.6]
** ———. *Finite Simple Groups: An Introduction to Their Classification.* [13.5]
* ———. *The Classification of Finite Simple Groups.* [13.5]
Goritz, John. *Mathematics for Welding Trades.* [16.2]
Gorsline, G.W. *Computer Organization: Hardware/Software.* [22.17]
Gostiaux, Bernard. *Differential Geometry: Manifolds, Curves, and Surfaces.* [14.10]
** Gould, Ronald. *Graph Theory.* [10.5]
Gould, Stephen Jay. *The Mismeasure of Man.* [25.1]
* Gould, Sydney H. *Russian for the Mathematician.* [2.2]
* ———. *Variational Methods for Eigenvalue Problems: An Introduction to the Method.* [7.6]
Goulden, I.P. *Combinatorial Enumeration.* [10.4]
Gowar, Norman. *An Invitation to Mathematics.* [1.4]
** Grabiner, Judith V. *The Origins of Cauchy's Rigorous Calculus.* [3.9]
Gradshteyn, I.S. *Table of Integrals, Series, and Products.* [2.2]
* Graham, L.A. *Ingenious Mathematical Problems and Methods.* [4.4]
———. *The Surprise Attack in Mathematical Problems.* [4.4]
Graham, Neill. *The Many-Worlds Interpretation of Quantum Mechanics.* [24.5]
*** Graham, Ronald L. *Concrete Mathematics: A Foundation for Computer Science.* [10.3]
** ———. *Ramsey Theory.* [10.4]
* ———. *Old and New Problems and Results in Combinatorial Number Theory.* [11.3]
* Gramain, André. *Topology of Surfaces.* [15.3]
** Grandine, Thomas A. *The Numerical Methods Programming Projects Book.* [18.6]
* Grandmont, Jean-Michel. *Temporary Equilibrium: Selected Readings.* [25.2]
Grant, Edward. *A Source Book in Medieval Science.* [3.3]
Grant, John. *Logical Introduction to Databases.* [22.5]
* Grassl, Richard M. *Discrete and Combinatorial Mathematics.* [10.3]
* Grattan-Guinness, I. *From the Calculus to Set Theory, 1630–1910: An Introductory History.* [3.9]
———. *History in Mathematics Education.* [5.5]
———. *Joseph Fourier, 1768–1830.* [3.2]
Grätzer, George. *General Lattice Theory.* [10.7]
* ———. *Universal Algebra.* [13.12]
** Graubard, Stephen R. *The Artificial Intelligence Debate: False Starts, Real Foundations.* [22.2]
Grauert, H. *Several Complex Variables.* [8.7]
* Graver, Jack E. *Incidence and Symmetry in Design and Architecture.* [14.5]
*** Gray, Jeremy. *The History of Mathematics: A Reader.* [3.1]
** ———. *Ideas of Space: Euclidean, Non-Euclidean, and Relativistic.* [14.4]
** Gray, Robert M. *Probability, Random Processes, and Ergodic Properties.* [20.8]
Gray, Theodore W. *Exploring Mathematics with Mathematica: Dialogs Concerning Computers and Mathematics.* [22.1]
Graybeal, Wayne J. *Simulation: Principles and Methods.* [19.9]
Graybill, Franklin A. *Introduction to the Theory of Statistics.* [21.4]
———. *Introduction to Matrices with Applications in Statistics.* [12.3]
———. *Theory and Application of the Linear Model.* [21.9]
Grazda, Edward E. *Handbook of Applied Mathematics.* [2.2]
Green, A.E. *Theoretical Elasticity.* [24.4]
Green, Paul E. *Mathematical Tools for Applied Multivariate Analysis.* [21.10]
* Greenberg, Marvin Jay. *Lectures on Algebraic Topology.* [15.3]
* ———. *Euclidean and Non-Euclidean Geometries: Development and History.* [14.4]
Greenberg, Michael D. *Application of Green's Functions in Science and Engineering.* [7.5]
** ———. *Foundations of Applied Mathematics.* [24.2]
Greenblatt, M.H. *Mathematical Entertainments.* [4.6]
Greene, Daniel H. *Mathematics for the Analysis of Algorithms.* [10.4]
Greene, Robert E. *Introduction to Topology.* [15.1]
** Greenspan, Donald. *Numerical Analysis for Applied Mathematics, Science, and Engineering.* [18.2]
Greenwood, J. Arthur. *Guide to Tables in Mathematical Statistics.* [2.4]

H

———. *Finite Mathematics and Its Applications.* [10.2]

* Haggarty, Rod. *Alice in Numberland: A Students' Guide to the Enjoyment of Higher Mathematics.* [1.4]

Haight, F.A. *Mathematical Theories of Traffic Flow.* [19.3]

* Haines, Robert G. *Math Principles for Food Service Occupations.* [16.2]

Hájek, Otomar. *Pursuit Games: An Introduction to the Theory and Applications of Differential Games of Pursuit and Evasion.* [19.4]

* Halberstam, H. *Sieve Methods.* [11.4]

Hald, Anders. *A History of Probability and Statistics and Their Applications Before 1750.* [3.11]

Hale, Jack K. *Studies in Ordinary Differential Equations.* [7.2]

———. *An Introduction to Infinite Dimensional Dynamical Systems—Geometric Theory.* [7.3]

———. *Methods of Bifurcation Theory.* [7.3]

* ———. *Ordinary Differential Equations.* [7.2]

Hall, A. Rupert. *Philosophers at War: The Quarrel Between Newton and Leibniz.* [3.9]

Hall, G. *Modern Numerical Methods for Ordinary Differential Equations.* [18.3]

Hall, Marshall, Jr. *Groups of Order* $2^n (n \leq 6)$. [13.5]

* ———. *Combinatorial Theory.* [10.4]

*** ———. *The Theory of Groups.* [13.5]

** Hallam, Thomas G. *Mathematical Ecology: An Introduction.* [23.2]

** ———. *Applied Mathematical Ecology.* [23.2]

Halmos, Paul R. *Reviews in Operator Theory, 1980–1986.* [2.6]

———. *A Hilbert Space Problem Book.* [8.8]

*** ———. *Finite-Dimensional Vector Spaces.* [12.2]

———. *I Have A Photographic Memory.* [3.6]

* ———. *I Want To Be A Mathematician: An Automathography in Three Parts.* [3.2]

———. *Introduction to Hilbert Space.* [8.8]

* ———. *Lectures on Boolean Algebras.* [13.13]

** ———. *Measure Theory.* [8.3]

*** ———. *Naive Set Theory.* [9.5]

** Halstead, Robert H., Jr. *Computation Structures.* [22.17]

Halsted, George B. *Girolamo Saccheri's Euclides Vindicatus.* [3.4]

* Hamilton, L.F. *Calculations of Analytic Chemistry.* [16.6]

Hammer, J. *Unsolved Problems Concerning Lattice Points.* [14.12]

Hammersley, J.M. *Monte Carlo Methods.* [21.8]

** Hamming, Richard W. *Coding and Information Theory.* [10.6]

* ———. *Introduction to Applied Numerical Analysis.* [18.1]

* ———. *Methods of Mathematics Applied to Calculus, Probability, and Statistics.* [6.3]

Hampel, Frank R. *Robust Statistics: The Approach Based on Influence Functions.* [21.5]

Hanan, James. *Lindenmayer Systems, Fractals, and Plants.* [23.2]

Hand, D.J. *The Statistical Consultant in Action.* [21.1]

———. *Discrimination and Classification.* [21.10]

Handelman, G.H. *Mathematics Applied to Continuum Mechanics.* [24.9]

Handscomb, D.C. *Monte Carlo Methods.* [21.8]

Hankins, Thomas L. *Sir William Rowan Hamilton.* [3.2]

* Hanks, Robert. *A Modern Course on the Theory of Equations.* [13.2]

Hanna, J. Ray. *Fourier Series and Integrals of Boundary Value Problems.* [7.5]

Hansen, Viggo P. *Computers in Mathematics Education: 1984 Yearbook.* [5.14]

Hao, Bai-Lin. *Chaos.* [7.3]

Harary, Frank. *Distance in Graphs.* [10.5]

———. *Graph Theory.* [10.5]

Harbison, Samuel P. *C: A Reference Manual.* [22.7]

*** Hardy, G.H. *Introduction to the Theory of Numbers.* [11.3]

*** ———. *Inequalities.* [8.11]

*** ———. *A Course of Pure Mathematics.* [6.4]

*** ———. *A Mathematician's Apology.* [1.2]

———. *Divergent Series.* [8.3]

* ———. *Ramanujan.* [3.2]

* Hardy, Kenneth. *The Green Book: 100 Practice Problems for Undergraduate Mathematics Competitions.* [4.4]

* Harel, David. *Algorithmics: The Spirit of Computing.* [22.1]

Hargittai, Istvan. *Symmetry: Unifying Human Understanding.* [14.5]

* Harper, John R. *Lectures on Algebraic Topology.* [15.3]

Harper, W.L. *Foundations of Probability Theory, Statistical Inference, and Statistical Theories of Sciences.* [20.5]

*** Harris, Carl M. *Fundamentals of Queueing Theory.* [19.8]

Harris, Joseph. *Principles of Algebraic Geometry.* [14.11]

Harris, Mary Dee. *Introduction to Natural Language Processing.* [22.11]

* Harrison, Michael A. *Introduction to Formal Language Theory.* [22.9]

* Hart, Derek. *Modelling with Projectiles.* [24.13]

** Hart, Sergiu. *Handbook of Game Theory with Applications to Economics.* [19.4]

Hart, Therese A. *A Challenge of Numbers: People in the Mathematical Sciences.* [5.1]

Hart, W.L. *Mathematics of Investment.* [17.5]

Hartley, H.O. *Guide to Tables in Mathematical Statistics.* [2.4]

Hartley, Roger. *Linear and Nonlinear Programming: An Introduction to Linear Methods in Mathematical Programming.* [19.5]

* Hartman, Philip. *Ordinary Differential Equations.* [7.2]

* Hartmanis, Juris. *Computational Complexity Theory.* [22.9]

Hartsfield, Nora. *Pearls in Graph Theory: A Comprehensive Introduction.* [10.5]

* Hill, Raymond. *A First Course in Coding Theory.* [10.6]

** Hill, R. *The Mathematical Theory of Plasticity.* [24.4]

Hill, Victor E. *Groups, Representations, and Characters.* [13.5]

* Hille, Einar. *Functional Analysis and Semi-Groups.* [8.8]

* ———. *Analytic Function Theory.* [8.7]

———. *Ordinary Differential Equations in the Complex Domain.* [7.2]

*** Hillier, Frederick S. *Introduction to Operations Research.* [19.1]

* Hillman, Abraham P. *Discrete and Combinatorial Mathematics.* [10.3]

———. *A First Undergraduate Course in Abstract Algebra.* [13.1]

* Hilton, Peter J. *Build Your Own Polyhedra.* [14.5]

* ———. *A Course in Homological Algebra.* [13.9]

Hindley, J. Roger. *Introduction to Combinators and λ-Calculus.* [9.9]

** Hinkley, David V. *Statistical Theory and Modelling.* [21.5]

———. *Theoretical Statistics.* [21.5]

Hintikka, Jaakko. *The Philosophy of Mathematics.* [9.4]

Hiorns, R.W. *The Mathematical Theory of the Dynamics of Biological Populations.* [23.5]

** Hirsch, Christian R. *The Secondary School Mathematics Curriculum: 1985 Yearbook.* [5.8]

*** Hirsch, Morris W. *Differential Equations, Dynamical Systems, and Linear Algebra.* [7.2]

* ———. *Differential Topology.* [15.4]

Hirschfeld, J.W.P. *Finite Projective Spaces of Three Dimensions.* [14.7]

———. *Projective Geometries Over Finite Fields.* [14.7]

** Hirshfield, Stuart. *The Analytical Engine.* [22.1]

Hirst, Ann. *Proceedings of the Sixth International Congress on Mathematical Education.* [5.6]

Hirst, Keith. *Proceedings of the Sixth International Congress on Mathematical Education.* [5.6]

* Hoaglin, David C. *Exploring Data Tables, Trends, and Shapes.* [21.6]

*** ———. *Understanding Robust and Exploratory Data Analysis.* [21.6]

———. *Applications, Basics, and Computing of Exploratory Data Analysis.* [21.6]

Hoare, C.A.R. *Communicating Sequential Processes.* [22.10]

———. *Structured Programming.* [22.6]

Hobbs, Margie. *Introduction to Technical Mathematics.* [16.1]

Hochstadt, Harry. *Differential Equations.* [7.1]

———. *Integral Equations.* [7.6]

** ———. *The Functions of Mathematical Physics.* [24.1]

** Hodges, Andrew. *Alan Turing, The Enigma.* [3.2]

Hodges, Joseph L. *Elements of Finite Probability.* [20.2]

* Hodges, W. *Building Models by Games.* [9.7]

Hodson, F.R. *Mathematics and Computers in Archaeology.* [25.1]

Hoel, Paul G. *Introduction to Probability Theory.* [20.2]

Hoenig, Alan. *Applied Finite Mathematics.* [10.2]

Hoffer, Alan. *Geometry.* [14.3]

Hoffman, Edward G. *Practical Problems in Mathematics for Machinists.* [16.2]

*** Hoffman, Kenneth. *Linear Algebra.* [12.2]

* ———. *Banach Spaces of Analytic Functions.* [8.8]

* Hoffman, Paul. *Archimedes' Revenge: The Joys and Perils of Mathematics.* [1.2]

* Hoffmann, Banesh. *Albert Einstein: Creator and Rebel.* [3.2]

Hoffmann, Laurence D. *Calculus for Business, Economics, and the Social and Life Sciences.* [17.4]

*** Hofstadter, Douglas R. *Gödel, Escher, Bach: An Eternal Golden Braid.* [1.2]

———. *Metamagical Themas: Questing for the Essence of Mind and Pattern.* [1.2]

Hogben, Lancelot. *Mathematics for the Million.* [1.2]

Hogg, Robert V. *Introduction to Mathematical Statistics.* [21.4]

* ———. *Loss Distributions.* [17.7]

———. *Engineering Statistics.* [21.4]

** ———. *Probability and Statistical Inference.* [21.4]

———. *Studies in Statistics.* [21.1]

Hogger, Christopher J. *Introduction to Logic Programming.* [22.6]

* Hohn, Franz E. *Applied Modern Algebra.* [13.3]

** Hoit, Laura K. *The Arithmetic of Dosages and Solutions.* [16.3]

Holden, Alan. *Orderly Tangles: Cloverleafs, Gordian Knots, and Regular Polylinks.* [14.5]

* ———. *Shapes, Space, and Symmetry.* [14.5]

Hollander, Myles. *The Statistical Exorcist: Dispelling Statistics Anxiety.* [21.1]

* ———. *Statistics: A Biomedical Introduction.* [21.3]

Hollingdale, Stuart. *Makers of Mathematics.* [3.1]

Holmes, Mark H. *Mathematical Modeling of the Hearing Process.* [23.6]

Holmes, Peter. *The Best of "Teaching Statistics."* [5.8]

*** Holmes, Philip. *Nonlinear Oscillations, Dynamical Systems, and Bifurcations of Vector Fields.* [7.3]

* Honsberger, Ross. *Ingenuity in Mathematics.* [1.2]

** ———. *Mathematical Gems.* [1.2]

* ———. *Mathematical Morsels.* [1.2]

* ———. *Mathematical Plums.* [1.2]

* ———. *More Mathematical Morsels.* [1.2]

Hooke, Robert. *How to Tell the Liars from the Statisticians.* [21.1]

Hooker, C.A. *Foundations of Probability Theory, Statistical Inference, and Statistical Theories of Sciences.* [20.5]

*** Hopcroft, John E. *Data Structures and Algorithms.* [22.4]

*** ———. *The Design and Analysis of Computer Algorithms.* [22.8]

*** ———. *Introduction to Automata Theory, Languages, and Computation.* [22.15]

Hopf, Heinz. *Differential Geometry in the Large.* [14.10]

I

J

K

Kansy, K. *Computer Graphics Programming: GKS—The Graphics Standard.* [22.18]
* Kao, T.I. *Was Pythagoras Chinese? An Examination of Right Triangle Theory in Ancient China.* [3.5]
Kapadia, C.H. *Handbook of Statistical Distributions.* [2.2]
Kaplan, Edward L. *Mathematical Programming and Games.* [19.5]
* Kaplan, Wilfred. *Advanced Calculus.* [6.4]
* Kaplansky, Irving. *Commutative Rings.* [13.8]
** ———. *Fields and Rings.* [13.6]
* ———. *Infinite Abelian Groups.* [13.5]
* ———. *Lie Algebras and Locally Compact Groups.* [13.11]
* ———. *Linear Algebra and Geometry: A Second Course.* [12.2]
———. *Matters Mathematical.* [1.4]
* ———. *Set Theory and Metric Spaces.* [15.1]
** Kappraff, Jay. *Connections: The Geometric Bridge Between Art and Science.* [1.2]
Kapur, J.N. *Fascinating World of Mathematical Sciences.* [1.1]
———. *Mathematical Modelling.* [19.2]
Karlin, Samuel. *Theoretical Studies on Sex Ratio Evolution.* [23.4]
———. *Tchebycheff Systems.* [8.3]
* ———. *A First Course in Stochastic Processes.* [20.4]
———. *A Second Course in Stochastic Processes.* [20.4]
* ———. *An Introduction to Stochastic Modeling.* [20.4]
Karlqvist, Anders. *Complexity, Language, and Life: Mathematical Approaches.* [23.8]
Karpinski, Louis C. *The History of Arithmetic.* [3.7]
Karplus, Walter J. *Digital Computer Treatment of Partial Differential Equations.* [18.3]
* Kato, Tosio. *Perturbation Theory for Linear Operators.* [8.9]
Katzan, Harry. *Operating Systems: A Pragmatic Approach.* [22.10]
* Katznelson, Yitzhak. *An Introduction to Harmonic Analysis.* [8.12]
Kauffman, Louis H. *On Knots.* [15.2]
Kaufmann, Arnold. *Introduction to Fuzzy Arithmetic: Theory and Applications.* [20.7]
* ———. *Introduction to the Theory of Fuzzy Subsets.* [20.7]
* Kaufmann, Jerome E. *College Algebra and Trigonometry.* [6.2]
Kavanau, J. Lee. *Symmetry: An Analytical Treatment.* [14.5]
*** Kawada, Yukiyosi. *Encyclopedic Dictionary of Mathematics.* [2.3]
* Kazarinoff, Nicholas D. *Analytic Inequalities.* [8.11]
** ———. *Geometric Inequalities.* [14.9]
* Kee, Joyce L. *Clinical Calculations with Applications to General and Specialty Areas.* [16.3]
* Keedwell, A.D. *Latin Squares and Their Applications.* [10.4]
Keen, Linda. *Chaos and Fractals: The Mathematics Behind the Computer Graphics.* [8.5]
Keener, James P. *Principles of Applied Mathematics: Transformation and Approximation.* [24.2]
* Keisler, H. Jerome. *Elementary Calculus.* [6.3]
* ———. *Foundations of Infinitesimal Calculus.* [6.3]
*** ———. *Model Theory.* [9.7]
Keitel, Christine. *Curriculum Development in Mathematics.* [5.6]
* Keith, Sandra. *Winning Women Into Mathematics.* [5.10]
* Keller, H.B. *Numerical Methods for Two-Point Boundary-Value Problems.* [18.3]
* Kelley, John L. *General Topology.* [15.1]
** Kellison, Stephen G. *The Theory of Interest.* [17.5]
** Kelly, L.M. *The William Lowell Putnam Mathematical Competition: Problems and Solutions, 1938–1964.* [4.4]
Kelly, Paul J. *The Non-Euclidean, Hyperbolic Plane: Its Structure and Consistency.* [14.4]
———. *Geometry and Convexity: A Study in Mathematical Methods.* [14.9]
Kelton, W.D. *Simulation Modeling and Analysis.* [19.9]
Kemeny, John G. *Finite Markov Chains.* [20.4]
———. *Mathematical Models in the Social Sciences.* [25.1]
———. *Denumerable Markov Chains.* [20.4]
*** ———. *Introduction to Finite Mathematics.* [10.2]
———. *Man and the Computer.* [22.2]
Kemp, Rainer. *Fundamentals of the Average Case Analysis of Particular Algorithms.* [22.8]
Kempe, A.B. *How to Draw a Straight Line.* [14.3]
Kempthorne, Oscar. *Probability, Statistics, and Data Analysis.* [21.4]
Kendall, Maurice. *Bibliography of Statistical Literature.* [2.4]
*** ———. *The Advanced Theory of Statistics.* [21.5]
———. *Multivariate Analysis.* [21.10]
* Kendig, Keith. *Elementary Algebraic Geometry.* [14.11]
Kennedy, Hubert C. *Peano: Life and Works of Giuseppe Peano.* [3.2]
* Kennedy, Peter. *A Guide to Econometrics.* [25.2]
Kennedy, William J., Jr. *Statistical Computing.* [21.8]
* Kenschaft, Patricia C. *Winning Women Into Mathematics.* [5.10]
* Kernighan, Brian W. *The UNIX Programming Environment.* [22.10]
*** ———. *The Elements of Programming Style.* [22.6]
*** ———. *The C Programming Language.* [22.7]
———. *The AWK Programming Language.* [22.10]
*** Keyfitz, Nathan. *Demography Through Problems.* [23.5]
* ———. *Applied Mathematical Demography.* [23.5]
———. *Introduction to the Mathematics of Population.* [23.5]
** ———. *Mathematical Demography.* [23.5]
* Kfoury, A.J. *An Introduction to Formal Language Theory.* [22.9]
Khalil, Hatem M. *Principles of Computer Science: Concepts, Algorithms, Data Structures, and Applications.* [22.3]

** Koblitz, Ann Hibner. *A Convergence of Lives, Sofia Kovalevskaia: Scientist, Writer, Revolutionary.* [3.2]
Koblitz, Neal. *A Course in Number Theory and Cryptography.* [11.3]
* ———. *Introduction to Elliptic Curves and Modular Forms.* [11.7]
———. *p-adic Analysis: A Short Course on Recent Work.* [11.8]
———. *p-adic Numbers, p-adic Analysis, and Zeta-Functions.* [11.8]
Koch, Helmut. *Introduction to Classical Mathematics I: From the Quadratic Reciprocity Law to the Uniformization Theorem.* [1.3]
Koehler, Vera J. *Programmed Mathematics of Drugs and Solutions.* [16.3]
Koffman, Elliot B. *Pascal: Problem Solving and Program Design.* [22.7]
Kogan, B. Yu. *The Application of Mechanics to Geometry.* [14.4]
Kohn, J.J. *Reviews in Complex Analysis, 1980–1986.* [2.6]
Kolatis, Maria S. *Mathematics for Data Processing and Computing.* [16.4]
* Kolman, Bernard. *Elementary Linear Programming with Applications.* [19.5]
* ———. *Applied Finite Mathematics.* [10.2]
Kolmogorov, Andrei N. *Introductory Real Analysis.* [8.2]
** ———. *Mathematics: Its Content, Methods, and Meaning.* [1.1]
** ———. *Foundations of the Theory of Probability.* [20.5]
Konheim, Alan G. *Cryptography: A Primer.* [10.6]
Kõnig, Dénes. *Theory of Finite and Infinite Graphs.* [10.5]
* Konkle, Gail S. *Shapes and Perceptions: An Intuitive Approach to Geometry.* [14.3]
** Koopmans, Lambert H. *Introduction to Contemporary Statistical Methods.* [21.2]
Kopchenova, N.V. *Computational Mathematics: Worked Examples and Problems with Elements of Theory.* [18.6]
Kopperman, Ralph. *Model Theory and Its Applications.* [9.7]
Koptsik, V.A. *Symmetry in Science and Art.* [14.5]
Kordemsky, Boris A. *The Moscow Puzzles: 359 Mathematical Recreations.* [4.3]
Kõrner, Stephan. *The Philosophy of Mathematics: An Introductory Essay.* [9.4]
* Kõrner, T.W. *Fourier Analysis.* [8.4]
Korovkin, P.P. *Inequalities.* [8.11]
* Korsh, James F. *Data Structures, Algorithms, and Program Style Using C.* [22.4]
Kortanek, Kenneth O. *Extremal Methods of Operations Research.* [19.1]
Korth, Henry F. *Database System Concepts.* [22.5]
* Kosko, Bart. *Neural Networks and Fuzzy Systems.* [22.11]
* Kośniowski, Czes. *A First Course in Algebraic Topology.* [15.3]
* ———. *Puzzle It Out: Cubes, Groups and Puzzles.* [4.2]

Kostrikin, A.I. *Algebra I: Basic Notions of Algebra.* [13.4]
———. *Algebra II: Non-Commutative Rings, Identities.* [13.6]
———. *Introduction to Algebra.* [13.1]
** Kotz, Samuel. *Encyclopedia of Statistical Sciences.* [2.3]
** ———. *Continuous Univariate Distributions.* [21.5]
** ———. *Discrete Distributions.* [21.5]
———. *Urn Models and Their Application: An Approach to Modern Discrete Probability Theory.* [20.3]
* Kovach, Ladis D. *Advanced Engineering Mathematics.* [24.2]
Kra, Irwin. *Reviews in Complex Analysis, 1980–1986.* [2.6]
*** Kraitchik, Maurice. *Mathematical Recreations.* [4.1]
Krakowiak, Sacha. *Principles of Operating Systems.* [22.10]
** Kramer, Arthur D. *Fundamentals of Technical Mathematics with Calculus.* [16.7]
** Kramer, Edna E. *The Nature and Growth of Modern Mathematics.* [3.1]
*** Krantz, Steven G. *Complex Analysis: The Geometric Viewpoint.* [8.7]
Krasnov, M.L. *Problems and Exercises in the Calculus of Variations.* [8.10]
** Krause, Eugene F. *Taxicab Geometry: An Adventure in Non-Euclidean Geometry.* [14.1]
Kreyszig, Erwin. *Advanced Engineering Mathematics.* [24.2]
* Krieger, Henry A. *Measure-theoretic Probability.* [20.6]
Kritikos, Nikolaos. *Lectures on Number Theory.* [11.3]
Krüger, Lorenz. *The Probabilistic Revolution.* [3.11]
* Kruse, Robert L. *Data Structures and Program Design.* [22.4]
Kruskal, William H. *International Encyclopedia of Statistics.* [2.3]
Kuchemann, D. *A Review of Research in Mathematical Education: Research on Learning and Teaching.* [5.12]
Kuh, Edwin. *Regression Diagnostics: Identifying Influential Data and Sources of Collinearity.* [21.9]
* Kuhfittig, Peter K.F. *Basic Technical Mathematics with Calculus.* [16.7]
———. *Introduction to Technical Mathematics.* [16.1]
Kulm, Gerald. *Assessing Higher Order Thinking in Mathematics.* [5.13]
Kumar, Vipin. *Search in Artificial Intelligence.* [22.11]
* Kumpel, P.G. *Linear Algebra with Applications to Differential Equations.* [12.1]
** Kunen, Kenneth. *Set Theory: An Introduction to Independence Proofs.* [9.5]
Kunz, E. *Introduction to Commutative Algebra and Algebraic Geometry.* [13.8]
*** Kunze, Ray. *Linear Algebra.* [12.2]
Kupper, L.L. *Epidemiologic Research: Principles and Quantitative Methods.* [23.3]

L

* Lazarsfeld, Paul F. *Mathematical Thinking in the Social Sciences.* [25.1]

* Lazou, Christopher. *Supercomputers and their Use.* [22.17]

Le Lionnais, F. *Great Currents of Mathematical Thought.* [1.1]

Leadbetter, M.R. *Stationary and Related Stochastic Processes.* [20.4]

Lebesgue, Henri. *Measure and the Integral.* [3.9]

*** LeBlanc, Richard J., Jr. *Crafting a Compiler.* [22.16]

** Leder, Gilah. *Mathematics and Gender.* [5.10]

Ledermann, Walter. *Algebra.* [13.4]

———. *Analysis.* [24.1]

* ———. *Geometry and Combinatorics.* [14.9]

** ———. *Introduction to Group Characters.* [13.5]

Ledolter, Johannes. *Engineering Statistics.* [21.4]

* Lee, Shin-Ying. *Mathematical Knowledge of Japanese, Chinese, and American Elementary School Children.* [5.13]

* LeFevor, C.S. *Mathematics for Auto Mechanics.* [16.2]

Leffin, Walter W. *Introduction to Technical Mathematics.* [16.1]

Leffler, Samuel J. *The Design and Implementation of the 4.3BSD UNIX Operating System.* [22.10]

* Lefschetz, Solomon. *Differential Equations: Geometric Theory.* [7.2]

———. *Stability by Liapunov's Direct Method.* [7.3]

———. *Topology.* [15.3]

** Lehmann, E.L. *Nonparametrics: Statistical Methods Based on Ranks.* [21.11]

———. *Elements of Finite Probability.* [20.2]

———. *Theory of Point Estimation.* [21.5]

Lehto, Olli. *Proceedings of the International Congress of Mathematicians, Helsinki, 1978.* [2.7]

* Leinbach, L. Carl. *The Laboratory Approach to Teaching Calculus.* [5.9]

Leiserson, Charles E. *Introduction to Algorithms.* [22.8]

** Leithold, Louis. *Before Calculus: Functions, Graphs, and Analytic Geometry.* [6.2]

Leitmann, George. *The Calculus of Variations and Optimal Control: An Introduction.* [19.10]

*** Leitzel, James R.C. *A Call For Change: Recommendations for the Mathematical Preparation of Teachers of Mathematics.* [5.9]

Leitzel, Joan R. *Transition to College Mathematics.* [6.2]

Lenz, Hanfried. *Design Theory.* [10.7]

Leon, Steven J. *Linear Algebra with Applications.* [12.1]

* Leontief, W. *Input-Output Economics.* [25.2]

* Lerner, Norbert. *Algebra and Trigonometry: A Pre-Calculus Approach.* [6.2]

Lesh, R. *Acquisition of Mathematics Concepts and Processes.* [5.3]

*** Lesniak, Linda. *Graphs & Digraphs.* [10.5]

Lessard, Sabin. *Theoretical Studies on Sex Ratio Evolution.* [23.4]

Lester, Frank. *How To Evaluate Progress in Problem Solving.* [5.11]

Levasseur, Kenneth. *Applied Discrete Structures for Computer Science.* [10.1]

Levenbach, Hans. *The Professional Forecaster: The Forecasting Process Through Data Analysis.* [21.16]

LeVeque, William J. *Reviews in Number Theory, 1940–1972.* [2.6]

* ———. *Topics in Number Theory.* [11.3]

Levi, Howard. *Polynomials, Power Series, and Calculus.* [6.3]

Levin, B.R. *The Evolution of Sex.* [23.4]

* Levin, Bruce. *Statistics for Lawyers.* [21.1]

** Levin, Simon A. *Mathematical Ecology: An Introduction.* [23.2]

———. *Perspectives in Ecological Theory.* [23.2]

** ———. *Studies in Mathematical Biology.* [23.1]

** ———. *Applied Mathematical Ecology.* [23.2]

———. *Niche: Theory and Application.* [23.2]

*** Levinson, Norman. *Theory of Ordinary Differential Equations.* [7.2]

Levy, Azriel. *Basic Set Theory.* [9.5]

** ———. *Foundations of Set Theory.* [9.5]

** Levy, David N.L. *Computer Games I.* [22.11]

Lewis, E. *Network Models in Population Biology.* [23.5]

Lewis, P.A.W. *The Statistical Analysis of Series of Events.* [21.15]

Lewis, Philip G. *Approaching Precalculus Mathematics Discretely: Explorations in a Computer Environment.* [6.2]

* Lewontin, Richard C. *An Introduction to Genetic Analysis.* [23.4]

** Lĭ Yǎn. *Chinese Mathematics: A Concise History.* [3.5]

Lial, Margaret L. *Essentials of Geometry for College Students.* [14.2]

Liapunov, A. *Stability of Motion.* [7.3]

Libeskind, S. *Logo.* [14.8]

* Lidl, Rudolf. *Introduction to Finite Fields and Their Applications.* [13.7]

* ———. *Applied Abstract Algebra.* [13.3]

* Lieber, Lillian R. *Galois and the Theory of Groups: A Bright Star in Mathesis.* [13.7]

* ———. *The Einstein Theory of Relativity.* [24.7]

*** Lieberman, Gerald J. *Introduction to Operations Research.* [19.1]

Liebscher, Herbert. *Handbook of Mathematical Formulas.* [2.2]

** Lighthill, Michael James. *An Informal Introduction to Theoretical Fluid Mechanics.* [24.9]

** ———. *Mathematical Biofluiddynamics.* [23.6]

** ———. *Waves in Fluids.* [24.10]

Lightstone, A.H. *Nonarchimedean Fields and Asymptotic Expansions.* [9.7]

* ———. *Mathematical Logic: An Introduction to Model Theory.* [9.7]

———. *Symbolic Logic and the Real Number System: An Introduction to the Foundations of Number Systems.* [8.1]

Lilly, Susan C. *Pascal Plus Data Structures.* [22.4]

** Lin, C.C. *Mathematics Applied to Deterministic Problems in the Natural Sciences.* [24.3]

** Lindgren, Harry. *Geometric Dissections.* [4.6]

M

Mackey, George W. *The Mathematical Foundations of Quantum Mechanics.* [24.5]

* Mackey, M.C. *From Clocks to Chaos: The Rhythms of Life.* [23.6]

** Mackiw, George. *Applications of Abstract Algebra.* [13.3]

* MacLennan, Bruce J. *Functional Programming.* [22.6]

———. *Principles of Programming Languages: Design, Evaluation, and Implementation.* [22.6]

* Macmillan, R.H. *Geometric Symmetry.* [14.5]

** MacWilliams, F.J. *The Theory of Error-Correcting Codes.* [10.6]

Madachy, Joseph S. *Mathematics on Vacation.* [4.6]

* Madansky, Albert. *Prescriptions for Working Statisticians.* [21.4]

Madison, Bernard L. *A Challenge of Numbers: People in the Mathematical Sciences.* [5.1]

Magalhaes, Luis T. *An Introduction to Infinite Dimensional Dynamical Systems—Geometric Theory.* [7.3]

* Magid, Andy R. *Applied Matrix Models: A Second Course in Linear Algebra with Computer Applications.* [18.4]

Magidson, Susan. *The Mystical Machine: Issues and Ideas in Computing.* [22.1]

Magnus, Jan R. *Matrix Differential Calculus with Applications in Statistics and Econometrics.* [6.4]

Magnus, Wilhelm. *The History of Combinatorial Group Theory: A Case Study in the History of Ideas.* [3.8]

Magurn, Bruce A. *Reviews in K-Theory, 1940–1984.* [2.6]

Mahoney, Michael S. *The Mathematical Career of Pierre de Fermat, 1601–1665.* [3.2]

* Maistrov, Leonid E. *Probability Theory: A Historical Sketch.* [3.11]

Makarenko, G.I. *Problems and Exercises in the Calculus of Variations.* [8.10]

Makarov, V.L. *Mathematical Theory of Economic Dynamics and Equilibria.* [25.2]

* Maki, Daniel P. *Finite Mathematics.* [10.2]

———. *Mathematical Models and Applications with Emphasis on the Social, Life, and Management Sciences.* [25.1]

* Malcolm, Michael A. *Computer Methods for Mathematical Computations.* [18.6]

Malkevitch, Joseph. *Graphs, Models, and Finite Mathematics.* [10.2]

Mallios, William S. *Statistical Modeling: Applications in Contemporary Issues.* [21.2]

** Manber, Udi. *Introduction to Algorithms: A Creative Approach.* [22.8]

*** Mandelbrot, Benoit. *The Fractal Geometry of Nature.* [8.5]

** Mangel, Marc. *Dynamic Modeling in Behavioral Ecology.* [23.2]

Manheim, Jerome H. *The Genesis of Point Set Topology.* [3.9]

Mann, W. Robert. *Advanced Calculus.* [6.4]

Manoukian, Edward B. *Modern Concepts and Theorems of Mathematical Statistics.* [21.4]

Manvel, B. *Introduction to Combinatorial Theory.* [10.3]

March, Lionel. *The Geometry of Environment: An Introduction to Spatial Organization in Design.* [14.1]

Marcowitz, Ulrich. *Numerical Solution of Partial Differential Equations.* [18.3]

Marcus, Daniel A. *Number Fields.* [11.5]

* Marcus, Marvin. *Survey of Matrix Theory and Matrix Inequalities.* [12.3]

———. *Introduction to Modern Algebra.* [13.1]

** Marcus-Roberts, Helen. *Life Science Models.* [23.1]

Margolis, Emil J. *Chemical Principles in Calculations of Tonic Equilibria.* [16.6]

Margolus, Norman. *Cellular Automata Machines: A New Environment for Modeling.* [22.15]

Markushevich, A.I. *The Theory of Analytic Functions: A Brief Course.* [8.6]

Marlow, W.H. *Mathematics for Operations Research.* [19.1]

Marmo, Giuseppe. *Dynamical Systems: A Differential Geometric Approach to Symmetry and Reduction.* [14.10]

Maron, I.A. *Computational Mathematics: Worked Examples and Problems with Elements of Theory.* [18.6]

Maron, Melvin J. *Numerical Analysis: A Practical Approach.* [18.1]

Marsden, Jerrold E. *Manifolds, Tensor Analysis, and Applications.* [14.10]

** ———. *Vector Calculus.* [6.4]

* ———. *Calculus III.* [6.4]

* ———. *Calculus.* [6.3]

———. *A Mathematical Introduction to Fluid Mechanics.* [24.9]

* ———. *The Hopf Bifurcation and Its Applications.* [7.3]

* Marshal, Sally M. *Clinical Calculations with Applications to General and Specialty Areas.* [16.3]

Marshall, Albert W. *Inequalities: Theory of Majorization and Its Applications.* [8.11]

Martin, Donald A. *Axiomatic Set Theory.* [9.5]

* Martin, Edward. *Elements of Mathematics, Book B: Problem Book.* [6.1]

* Martin, George E. *The Foundations of Geometry and the Non-Euclidean Plane.* [14.4]

* ———. *Transformation Geometry: An Introduction to Symmetry.* [14.5]

Martin, Robert H., Jr. *Nonlinear Operators and Differential Equations in Banach Spaces.* [7.6]

* Marx, Morris L. *An Introduction to Mathematical Statistics and Its Applications.* [21.4]

* Masani, Pesi. *Norbert Wiener, 1894–1964.* [3.2]

* Mason, John H. *Thinking Mathematically.* [5.11]

* ———. *Learning and Doing Mathematics.* [5.11]

Mason, R.D. *Mathematics for Business and New Consumers.* [17.1]

* Mason, Robert L. *Statistical Design and Analysis of Experiments: With Applications to Engineering and Science.* [21.12]

———. *Regression Analysis and Its Application: A Data-Oriented Approach.* [21.9]

* ———. *Companion to Concrete Mathematics.* [4.1]
* ———. *Invitation to Geometry.* [14.2]
———. *Mathematical Ideas, Modeling, and Applications.* [19.2]
* Mendelson, Elliott. *Introduction to Mathematical Logic.* [9.3]
———. *Schaum's 3000 Solved Problems in Calculus.* [6.5]
Mendenhall, William. *A Second Course in Business Statistics: Regression Analysis.* [17.3]
** ———. *Elementary Survey Sampling.* [21.13]
———. *Mathematical Statistics with Applications.* [21.4]
* Menninger, Karl. *Number Words and Number Symbols: A Cultural History of Numbers.* [3.7]
** Merzbach, Uta C. *A Century of Mathematics in America.* [3.6]
*** ———. *A History of Mathematics.* [3.1]
Meserve, Bruce E. *Contemporary Mathematics.* [1.4]
Messick, David M. *Mathematical Thinking in Behavioral Sciences.* [25.4]
Mesterton-Gibbons, Michael. *A Concrete Approach to Mathematical Modelling.* [19.2]
Mestre, Jose P. *Academic Preparation In Science: Teaching for Transition From High School To College.* [5.1]
———. *Linguistic and Cultural Influences on Learning Mathematics.* [5.4]
Metropolis, N. *A History of Computing in the Twentieth Century: A Collection of Essays.* [3.12]
Metz, J.A.J. *The Dynamics of Physiologically Structured Populations.* [23.5]
Meyer, Bertrand. *Introduction to the Theory of Programming Languages.* [22.6]
** ———. *Object-Oriented Software Construction.* [22.6]
* Meyer, C.D., Jr. *Generalized Inverses of Linear Transformations.* [12.4]
* Meyer, Paul-André. *Probabilities and Potential.* [20.4]
Meyer, Richard E. *Introduction to Mathematical Fluid Dynamics.* [24.9]
Meyer, Richard M. *Essential Mathematics for Applied Fields.* [24.2]
* Meyer, Walter J. *Concepts of Mathematical Modeling.* [19.2]
Meyer, Walter. *Graphs, Models, and Finite Mathematics.* [10.2]
* Michaels, Brenda. *Calculus: Readings from the Mathematics Teacher.* [6.5]
Michels, Leo. *Mathematics for Health Sciences.* [16.3]
Michod, R.E. *The Evolution of Sex.* [23.4]
* Mickens, Ronald E. *An Introduction to Nonlinear Oscillations.* [7.6]
———. *Difference Equations.* [10.7]
Midici, Geraldine Ann. *Drug Dosage Calculations: A Guide to Clinical Calculation.* [16.3]
Miertschin, S.L. *Technical Mathematics with Calculus.* [16.7]
Mikami, Yashio. *The Development of Mathematics in China and Japan.* [3.5]
Milenkovic, Milan. *Operating Systems: Concepts and Design.* [22.10]
Miller, Charles D. *Mathematical Ideas.* [1.4]
Miller, John D. *Men and Institutions in American Mathematics.* [3.6]
Miller, Nancy E. *File Structures Using Pascal.* [22.4]
Miller, Richard K. *Nonlinear Volterra Integral Equations.* [7.6]
———. *Ordinary Differential Equations.* [7.1]
Miller, Robert B. *Intermediate Business Statistics.* [17.3]
Miller, Rupert G., Jr. *Biostatistics Casebook.* [23.7]
* Millman, Richard S. *Elements of Differential Geometry.* [14.10]
* ———. *Geometry: A Metric Approach with Models.* [14.2]
* Milne-Thomson, L.M. *Theoretical Hydrodynamics.* [24.9]
Milnor, John W. *Characteristic Classes.* [15.3]
* ———. *Morse Theory.* [15.4]
———. *Singular Points of Complex Hypersurfaces.* [15.3]
** ———. *Topology from the Differentiable Viewpoint.* [15.4]
* Minc, Henryk. *Nonnegative Matrices.* [12.3]
* ———. *Survey of Matrix Theory and Matrix Inequalities.* [12.3]
Mines, Ray. *A Course in Constructive Algebra.* [13.2]
** Minsky, Marvin. *Perceptrons: An Introduction to Computational Geometry.* [22.11]
* ———. *Computation: Finite and Infinite Machines.* [22.15]
** Mirkin, Boris G. *Group Choice.* [25.3]
*** Misner, Charles W. *Gravitation.* [24.7]
** Mitrinović, Dragoslav S. *Recent Advances in Geometric Inequalities.* [14.9]
* Mizrahi, Abe. *Finite Mathematics with Applications for Business and Social Sciences.* [17.2]
Mock, G.D. *Analytic Geometry: Two and Three Dimensions.* [6.2]
* Mohamed, J.L. *Numerical Algorithms.* [18.6]
* ———. *Computational Methods for Integral Equations.* [18.7]
Mohr, Georg. *Compendium Euclidis Curiosi: Or Geometrical Operations.* [3.4]
Moise, Edwin E. *Geometry.* [14.3]
———. *Elementary Geometry from an Advanced Standpoint.* [14.4]
———. *Geometric Topology in Dimensions 2 and 3.* [15.2]
———. *Introductory Problem Courses in Analysis and Topology.* [8.3]
Moler, Cleve B. *Numerical Methods and Software.* [18.6]
* ———. *Computer Methods for Mathematical Computations.* [18.6]
———. *Computer Solution of Linear Algebraic Systems.* [18.4]
* ———. *Experiments in Computational Matrix Algebra.* [18.4]
* Moll, Robert N. *An Introduction to Formal Language Theory.* [22.9]

Murnane, Richard J. _Improving Indicators of the Quality of Science and Mathematics Education in Grades K–12._ [5.12]

Murphy, Dennis. _Merchandising Mathematics._ [16.2]

* Murray, J.D. _Lectures on Nonlinear-Differential-Equation Models in Biology._ [23.8]

*** ———. _Mathematical Biology._ [23.1]

Murray, Walter. _Practical Optimization._ [19.6]

Murty, Katta G. _Linear and Combinatorial Programming._ [19.5]

*** Murty, U.S.R. _Graph Theory with Applications._ [10.5]

** Myerson, Roger B. _Game Theory: Analysis of Conflict._ [19.4]

N

Naber, Gregory L. _Topological Methods in Euclidean Spaces._ [15.3]

Nachbin, Leopoldo. _Introduction to Functional Analysis: Banach Spaces and Differential Calculus._ [8.8]

* Nagata, Masayoshi. _Local Rings._ [13.8]

** Nagel, Ernest. _Gödel's Proof._ [9.3]

* Nagell, Trygve. _Introduction to Number Theory._ [11.1]

Nagpaul, S.R. _Basic Abstract Algebra._ [13.1]

** Nagy, Bela Sz. _Functional Analysis._ [8.8]

* Naimark, M.A. _Normed Rings._ [8.8]

Napier, John. _Rabdology._ [3.4]

Narasimhan, R. _Complex Analysis in One Variable._ [8.6]

Nash, Stephen. _Numerical Methods and Software._ [18.6]

National Council of Teachers of Mathematics. _A Sourcebook of Applications of School Mathematics._ [5.6]

———. _Cumulative Index, The Arithmetic Teacher: 1974–1983._ [2.5]

*** ———. _Curriculum and Evaluation Standards for School Mathematics._ [5.6]

*** ———. _Professional Standards for Teaching Mathematics._ [5.6]

* ———. _The Mathematics Teacher: Cumulative Indices._ [2.5]

*** National Research Council. _Everybody Counts: A Report to the Nation on the Future of Mathematics Education._ [5.1]

** ———. _Moving Beyond Myths: Revitalizing Undergraduate Mathematics._ [5.1]

* ———. _Renewing U.S. Mathematics: A Plan for the 1990s._ [5.1]

* ———. _Renewing U.S. Mathematics: Critical Resource for the Future._ [5.1]

Natterer, F. _The Mathematics of Computerized Tomography._ [23.7]

Nazareth, J.L. _Computer Solution of Linear Programs._ [19.5]

* Nehari, Zeev. _Conformal Mapping._ [8.7]

———. _Introduction to Complex Analysis._ [8.6]

Neill, Alistair. _Life Contingencies._ [17.7]

* Nelder, J.A. _Generalized Linear Models._ [21.9]

** Nelson, C. Robert. _Elementary Technical Mathematics._ [16.1]

* Nelson, Edward. _Radically Elementary Probability Theory._ [20.4]

* Nelson, R.D. _The Penguin Dictionary of Mathematics._ [2.1]

*** Nemhauser, George L. _Integer and Combinatorial Optimization._ [19.7]

———. _Integer Programming._ [19.7]

Nemytskii, V.V. _Qualitative Theory of Differential Equations._ [7.2]

Nesbitt, Cecil J. _Mathematics of Compound Interest._ [17.5]

Nesher, Pearla. _Mathematics and Cognition._ [5.3]

* Neter, John. _Applied Linear Statistical Models: Regression, Analysis of Variance, and Experimental Designs._ [21.9]

Neudecker, Heinz. _Matrix Differential Calculus with Applications in Statistics and Econometrics._ [6.4]

Neugebauer, O. _A History of Ancient Mathematical Astronomy._ [3.5]

———. _Mathematical Astronomy in Copernicus's De Revolutionibus._ [3.4]

* ———. _The Exact Sciences in Antiquity._ [3.5]

Neuts, Marcel. _Structured Stochastic Matrices of M/G/1 Type and Their Applications._ [19.8]

Nevanlinna, Rolf. _Introduction to Complex Analysis._ [8.6]

* Newbold, Paul. _Statistics for Business and Economics._ [17.3]

* Newell, Allen. _Computer Structures: Principles and Examples._ [22.17]

Newell, Virginia K. _Black Mathematicians and Their Works._ [5.10]

** Newman, Claire M. _Exploring Probability._ [20.1]

* Newman, Donald J. _A Problem Seminar._ [4.4]

———. _Complex Analysis._ [8.6]

———. _Polynomial Approximation._ [18.5]

** Newman, James R. _Gödel's Proof._ [9.3]

*** ———. _The World of Mathematics._ [1.1]

Newman, M.H.A. _Elements of the Topology of Plane Sets of Points._ [15.1]

Newman, Morris. _Integral Matrices._ [12.3]

*** Newton, Isaac. _Mathematical Principles of Natural Philosophy._ [3.4]

Ney, P.E. _Branching Processes._ [19.3]

Nicodemi, Olympia. _Discrete Mathematics: A Bridge to Computer Science and Advanced Mathematics._ [10.1]

Nicolas, Gregoier. _Exploring Complexity: An Introduction._ [19.2]

* Niederreiter, Harald. _Introduction to Finite Fields and Their Applications._ [13.7]

* Nielson, Gregory M. _Visualization in Scientific Computing._ [22.18]

** Nievergelt, Jurg. _Combinatorial Algorithms: Theory and Practice._ [22.8]

* Nievergelt, Yves. _Mathematics in Business Administration._ [17.2]

* Nijenhuis, Albert. _Combinatorial Algorithms for Computers and Calculators._ [22.8]

O

* Orr, Eleanor Wilson. *Twice As Less: Black English and the Performance of Black Students in Mathematics and Science.* [5.10]
Orszag, Steven. *Advanced Mathematical Methods for Scientists and Engineers.* [24.3]
* Ortega, James M. *Numerical Analysis: A Second Course.* [18.2]
Orsech, Grace. *Plane Algebraic Curves: An Introduction Via Valuations.* [14.11]
Orsech, Morris. *Plane Algebraic Curves: An Introduction Via Valuations.* [14.11]
Osen, L.M. *Women in Mathematics.* [5.10]
Oster, George F. *Caste and Ecology in the Social Insects.* [23.2]
** Ott, Lyman. *Elementary Survey Sampling.* [21.13]
** Owen, David R. *A First Course in the Mathematical Foundations of Thermodynamics.* [24.13]
Owen, D.B. *On the History of Statistics and Probability.* [3.11]
———. *Handbook of Statistical Distributions.* [2.2]
Owen, Guillermo. *Game Theory.* [19.4]
** Oxtoby, John C. *Measure and Category: A Survey of the Analogies between Topological and Measure Spaces.* [15.1]

P

Paatero, V. *Introduction to Complex Analysis.* [8.6]
*** Packel, Edward W. *The Mathematics of Games and Gambling.* [20.1]
Paelinck, J.H.P. *Bifurcation Analysis: Principles, Applications, and Synthesis.* [7.3]
Page, E.S. *An Introduction to Computational Combinatorics.* [10.3]
Page, Warren. *Two-Year College Mathematics Readings.* [1.6]
* Pakin, Sandra. *APL: The Language and Its Usage.* [22.7]
* Palissier, Carol. *Algorithmic Methods for Artificial Intelligence.* [22.11]
Palmer, Claude I. *Practical Mathematics.* [16.2]
* Palmer, Edgar M. *Graphical Evolution: An Introduction to the Theory of Random Graphs.* [10.5]
* Panjer, Harry E. *Actuarial Mathematics.* [17.8]
Pankratz, Alan. *Forecasting With Univariate Box-Jenkins Models: Concepts and Cases.* [21.16]
Panton, Arthur William. *The Theory of Equations with an Introduction to the Theory of Binary Algebraic Forms.* [13.1]
* Papadimitriou, Christos H. *Combinatorial Optimization: Algorithms and Complexity.* [19.7]
———. *The Theory of Database Concurrency Control.* [22.5]
** Papert, Seymour. *Mindstorms: Children, Computers, and Powerful Ideas.* [5.14]
** ———. *Perceptrons: An Introduction to Computational Geometry.* [22.11]
Pareigis, B. *Categories and Functors.* [13.10]
Parent, D.P. *Exercises in Number Theory.* [11.9]
* Parikh, Carol. *The Unreal Life of Oscar Zariski.* [3.2]
* Parker, George D. *Elements of Differential Geometry.* [14.10]
* ———. *Geometry: A Metric Approach with Models.* [14.2]
Parker, R. Gary. *Discrete Optimization.* [19.7]
Parker, Thomas S. *Practical Numerical Algorithms for Chaotic Systems.* [18.3]
* Parmenter, Michael M. *Theory of Interest and Life Contingencies with Pension Applications: A Problem Solving Approach.* [17.7]
Parsons, Torrence D. *Mathematical Methods in Finance and Economics.* [25.2]
** Parthasarathy, K.R. *Probability Measures on Metric Spaces.* [20.6]
Parzen, Emanuel. *Modern Probability Theory and Its Applications.* [20.3]
Pasahow, Edward. *Mathematics for Electronics.* [16.5]
Passman, Donald S. *The Algebraic Structure of Group Rings.* [13.6]
*** Patashnik, Oren. *Concrete Mathematics: A Foundation for Computer Science.* [10.3]
Patel, Jagdish K. *Handbook of Statistical Distributions.* [2.2]
Patterson, Richard R. *Projective Geometry and Its Applications to Computer Graphics.* [14.7]
Paul, Richard S. *Essentials of Technical Mathematics.* [16.1]
** Paulos, John Allen. *Beyond Numeracy: Ruminations of a Numbers Man.* [1.2]
———. *Innumeracy: Mathematical Illiteracy and Its Consequences.* [1.2]
* Pavlidis, Theo. *Algorithms for Graphics and Image Processing.* [22.18]
———. *Biological Oscillators: Their Mathematical Analysis.* [23.6]
** Payne, Joseph N. *Mathematics For the Young Child.* [5.7]
———. *Advanced Mathematics: A Preparation for Calculus.* [6.2]
Pearce, Peter. *Polyhedra Primer.* [14.5]
Pearce, Susan. *Polyhedra Primer.* [14.5]
Pearcy, Carl. *Introduction to Operator Theory I: Elements of Functional Analysis.* [8.9]
Pearson, Carl E. *Handbook of Applied Mathematics: Selected Results and Methods.* [24.1]
———. *Ordinary Differential Equations.* [7.2]
** ———. *Partial Differential Equations: Theory and Technique.* [7.4]
** Pečarić, J.E. *Recent Advances in Geometric Inequalities.* [14.9]
* Peck, Roxy. *Statistics: The Exploration and Analysis of Data.* [21.2]
* Pedersen, Jean. *Build Your Own Polyhedra.* [14.5]
Pedley, T.J. *Scale Effects in Animal Locomotion.* [23.2]
Pedoe, Dan. *Circles: A Mathematical View.* [14.4]
———. *Geometry: A Comprehensive Course.* [14.2]
* ———. *Geometry and the Visual Arts.* [14.2]
———. *The Gentle Art of Mathematics.* [1.2]
Pedrick, George. *A First Course in Functional Analysis.* [8.8]

Pontrjagin, Lev S. *General Topology I: Basic Concepts and Constructions, Dimension Theory.* [15.1]
———. *The Mathematical Theory of Optimal Processes.* [19.10]
———. *Foundations of Combinatorial Topology.* [15.3]
———. *Ordinary Differential Equations.* [7.2]
———. *Topological Groups.* [15.3]
Pooch, Udo W. *Simulation: Principles and Methods.* [19.9]
** Port, Dan. *Differential Equations: Theory and Applications.* [7.1]
Port, Sidney C. *Brownian Motion and Classical Potential Theory.* [20.4]
———. *Introduction to Probability Theory.* [20.2]
Porteous, Ian R. *Topological Geometry.* [14.12]
Porter, Richard D. *Mathematics and Its Applications to Management, Life, and Social Sciences With Finite and Discrete Mathematics.* [10.2]
* Porter, Stuart R. *Basic Technical Mathematics with Calculus.* [16.7]
Porter, Theodore M. *The Rise of Statistical Thinking, 1820–1900.* [3.11]
Posamentier, Alfred S. *Excursions in Advanced Euclidean Geometry.* [14.4]
Post, Thomas R. *Teaching Mathematics in Grades K–8: Research Based Methods.* [5.7]
Post, W.M. *Positive Feedback in Natural Systems.* [23.2]
Postill, Ronald. *The Sunday Times Book of Brain Teasers.* [4.3]
Postnikov, Mikhail. *Lectures in Geometry.* [14.2]
** Poston, Tim. *Catastrophe Theory and its Applications.* [19.2]
* ———. *Tensor Geometry: The Geometric Viewpoint and its Uses.* [14.10]
Pottage, John. *Geometrical Investigations: Illustrating the Art of Discovery in the Mathematical Field.* [14.1]
* Pour-El, Marian B. *Computability in Analysis and Physics.* [9.6]
Power, Thomas C. *Electronics Mathematics.* [16.5]
* Powers, David L. *Boundary Value Problems.* [7.5]
Prabhu, N.U. *Stochastic Storage Processes: Queues, Insurance Risk, and Dams.* [19.3]
Prather, Ronald. *Elements of Discrete Mathematics.* [10.1]
Pratt, Philip J. *Database Systems: Management and Design.* [22.5]
Prenowitz, Walter. *Join Geometries: A Theory of Convex Sets and Linear Geometry.* [14.9]
———. *Basic Concepts of Geometry.* [14.2]
Prenter, P.M. *Splines and Variational Methods.* [18.5]
* Preparata, Franco P. *Computational Geometry: An Introduction.* [14.8]
* Press, S. James. *Bayesian Statistics: Principles, Models, and Applications.* [21.7]
*** Press, William H. *Numerical Recipes: The Art of Scientific Computing.* [18.6]
* Pressman, R. *Software Engineering: A Practitioner's Approach.* [22.14]
* Prest, M.Y. *Numbers, Groups, and Codes.* [13.2]
Preston, Chris. *Iterates of Maps on an Interval.* [8.5]
Price, Bertram. *Regression Analysis by Example.* [21.9]
Price, G. Baley. *An Introduction to Multicomplex Spaces and Functions.* [8.7]
———. *Multivariable Analysis.* [6.4]
Price, Jack. *Changing School Mathematics: A Responsive Process.* [5.1]
*** Prichett, Gordon D. *Mathematics with Applications in Management and Economics.* [17.6]
Priestley, H.A. *Introduction to Lattices and Order.* [10.7]
Priestley, William M. *Calculus: An Historical Approach.* [6.3]
Prigogine, Ilya. *Exploring Complexity: An Introduction.* [19.2]
* Proschan, Frank. *Statistical Theory of Reliability and Life Testing.* [19.3]
———. *The Statistical Exorcist: Dispelling Statistics Anxiety.* [21.1]
Protter, Murray H. *A First Course in Real Analysis.* [8.2]
* ———. *Maximum Principles in Differential Equations.* [7.4]
———. *Reviews in Partial Differential Equations, 1980–1986.* [2.6]
Provine, W. *The Origins of Theoretical Population Genetics.* [23.4]
Prusinkiewicz, Przemyslaw. *Lindenmayer Systems, Fractals, and Plants.* [23.2]
Pugh, Anthony. *An Introduction to Tensegrity.* [14.12]
———. *Polyhedra: A Visual Approach.* [14.5]
Pugh, John R. *Inside Smalltalk.* [22.7]
Pujolle, G. *Introduction to Queueing Networks.* [19.8]
Pullman, N.J. *Matrix Theory and its Applications: Selected Topics.* [12.3]
* Putnam, Hilary. *Philosophy of Mathematics: Selected Readings.* [9.4]

Q

Quine, Willard Van Orman. *Methods of Logic.* [9.2]
* ———. *Philosophy of Logic.* [9.4]
———. *Pursuit of Truth.* [9.4]
———. *Set Theory and Its Logic.* [9.5]
———. *The Ways of Paradox and Other Essays.* [9.4]
* Quinney, Douglas. *An Introduction to the Numerical Solution of Differential Equations.* [18.3]

R

Rabinowitz, Philip. *A First Course in Numerical Analysis.* [18.1]
** ———. *Methods of Numerical Integration.* [18.1]
Rachev, S.T. *Mathematical Models for the Construction of Queueing Models.* [19.8]
* Rademacher, Hans. *The Enjoyment of Mathematics.* [1.2]

Richmond, A.E. *Calculus for Electronics.* [16.5]
* Richter, P.H. *The Beauty of Fractals: Images of Complex Dynamical Systems.* [8.5]
Richter-Dyn, Nira. *Stochastic Models in Biology.* [23.1]
Riesel, Hans. *Prime Numbers and Computer Methods for Factorization.* [11.4]
Riess, R. Dean. *Numerical Analysis.* [18.1]
** Riesz, Frigyes. *Functional Analysis.* [8.8]
* Rindler, Wolfgang. *Essential Relativity: Special, General, and Cosmological.* [24.7]
Riner, John. *Mathematics of Finance.* [17.5]
Ringel, Gerhard. *Pearls in Graph Theory: A Comprehensive Introduction.* [10.5]
* Ringrose, John R. *Fundamentals of the Theory of Operator Algebras.* [8.9]
Riordan, John. *An Introduction to Combinatorial Analysis.* [10.4]
———. *Combinatorial Identities.* [10.4]
* Ripley, Brian D. *Stochastic Simulation.* [19.9]
Rippon, P.J. *Microcomputers and Mathematics.* [22.1]
* Rising, Gerald R. *Unified Mathematics.* [6.1]
*** Ritchie, Dennis M. *The C Programming Language.* [22.7]
Rivest, Ronald L. *Introduction to Algorithms.* [22.8]
Rivlin, Theodore J. *An Introduction to the Approximation of Functions.* [18.5]
———. *Chebyshev Polynomials.* [8.15]
Roach, G.F. *Green's Functions.* [7.5]
Robbins, Herbert. *Great Expectations: The Theory of Optimal Stopping.* [20.4]
*** Robbins, H. *What is Mathematics?* [1.2]
Robbins, Omer, Jr. *Tonic Reactions and Equilibria.* [16.6]
* Roberts, A. Wayne. *Faces of Mathematics: An Introductory Course for College Students.* [1.4]
Roberts, Charles E., Jr. *Ordinary Differential Equations: A Computational Approach.* [7.1]
** Roberts, Fred S. *Discrete and System Models.* [19.2]
** ———. *Applied Combinatorics.* [10.3]
* ———. *Discrete Mathematical Models with Applications to Social, Biological, and Environmental Problems.* [19.2]
* ———. *Graph Theory and Its Applications to Problems of Society.* [25.1]
Roberts, Keith. *Mathematics for Health Sciences.* [16.3]
* Roberts, Paul M. *Mathematical Writing.* [5.9]
** Robertson, John S. *Differential Equations with Applications and Historical Notes.* [7.1]
* Robins, Gay. *The Rhind Mathematical Papyrus: An Ancient Egyptian Text.* [3.4]
Robinson, Abraham. *Complete Theories.* [9.7]
** ———. *Introduction to Model Theory and to the Metamathematics of Algebra.* [9.7]
———. *Nonarchimedean Fields and Asymptotic Expansions.* [9.7]
** ———. *Non-standard Analysis.* [8.14]
* ———. *Numbers and Ideals.* [13.6]

Robitaille, David F. *The IEA Study of Mathematics II: Contexts and Outcomes of School Mathematics.* [5.13]
** Robson, David. *Smalltalk-80: The Language.* [22.7]
* Robson, J.C. *Noncommutative Noetherian Rings.* [13.6]
Roch, Elaine. *Artificial Intelligence.* [22.11]
Rodi, Stephen B. *New Directions in Two-Year College Mathematics.* [5.9]
Rodman, L. *Invariant Subspaces of Matrices with Applications.* [12.2]
Rogers, David F. *Mathematical Elements for Computer Graphics.* [22.18]
*** Rogers, Hartley, Jr. *Theory of Recursive Functions and Effective Computability.* [9.6]
Rogerson, Alan. *Numbers and Infinity: A Historical Account of Mathematical Concepts.* [1.4]
Rogosinski, Werner. *Fourier Series.* [8.4]
* Roitman, Judith. *Introduction to Modern Set Theory.* [9.5]
Rokhlin, V.A. *Beginner's Course in Topology: Geometric Chapters.* [15.1]
** Rolfsen, Dale. *Knots and Links.* [15.2]
** Rollett, A.P. *Mathematical Models.* [14.5]
* Roman, Steven. *An Introduction to Discrete Mathematics.* [10.1]
———. *The Umbral Calculus.* [12.4]
** Romano, Joseph P. *Counterexamples in Probability and Statistics.* [20.8]
Romanova, M.A. *Topics in Mathematical Geology.* [24.13]
Romberg, Thomas A. *Addition and Subtraction: A Cognitive Perspective.* [5.12]
* ———. *Mathematics Assessment and Evaluation: Imperatives for Mathematics Educators.* [5.13]
———. *School Mathematics: Options for the 1990's.* [5.1]
* Rorres, Chris. *Applications of Linear Algebra.* [12.1]
Rose, H.E. *A Course in Number Theory.* [11.1]
Rose, Nicholas J. *Mathematical Maxims and Minims.* [1.5]
* Rosen, Joe. *Symmetry Discovered: Concepts and Applications in Nature and Science.* [14.5]
Rosen, Kenneth H. *Discrete Mathematics and its Applications.* [10.1]
———. *Elementary Number Theory and Its Applications.* [11.1]
** Rosen, Michael I. *A Classical Introduction to Modern Number Theory.* [11.5]
* Rosenblatt, Murray. *Studies in Probability Theory.* [20.3]
———. *Random Processes.* [20.4]
Rosenfeld, B.A. *Stereographic Projection.* [14.4]
* ———. *A History of Non-Euclidean Geometry: Evolution of the Concept of a Geometric Space.* [3.10]
Rosenlicht, Maxwell. *Introduction to Analysis.* [8.2]
Ross, Ian C. *Index to Statistics and Probability: Permuted Titles.* [2.5]
** Ross, Kenneth A. *Discrete Mathematics.* [10.1]
** ———. *Elementary Analysis: The Theory of Calculus.* [8.2]

S

** ———. *Fundamentals of Data Structures in Pascal.* [22.4]

Saidan, A.S. *The Arithmetic of Al-Uqlīdisī.* [3.4]

** Saks, S. *Analytic Functions.* [8.7]

** Salkind, Charles T. *Contest Problem Book.* [4.4]

Sally, Judith D. *Emmy Noether in Bryn Mawr.* [3.2]

Salomaa, Arto. *Computation and Automata.* [22.15]

———. *Jewels of Formal Language Theory.* [22.9]

Salpeter, Edwin E. *Quantum Mechanics for One- and Two-Electron Atoms.* [24.5]

Samarskii, A.A. *Equations of Mathematical Physics.* [7.6]

* Samelson, Hans. *Notes on Lie Algebras.* [13.11]

Sammons, Vivian O. *Blacks in Science and Medicine.* [5.10]

Samuel, Pierre. *Algebraic Theory of Numbers.* [11.5]

*** ———. *Commutative Algebra.* [13.8]

———. *Projective Geometry.* [14.7]

* Sanchez, David A. *Ordinary Differential Equations and Stability Theory: An Introduction.* [7.2]

———. *Topics in Ordinary Differential Equations.* [7.2]

Sanchis, Luis E. *Reflexive Structures: An Introduction to Computability Theory.* [22.9]

Sanderson, B.J. *Introduction to Piecewise-Linear Topology.* [15.2]

Sandler, Howard M. *Statistical Illusions: Problems.* [21.3]

Sankappanavar, H.P. *A Course in Universal Algebra.* [13.12]

Santaló, Luis A. *Integral Geometry and Geometric Probability.* [14.12]

Sarmanov, O.V. *Topics in Mathematical Geology.* [24.13]

Sassone, Peter G. *Cost-Benefit Analysis: A Handbook.* [25.2]

Sato, Ryuzo. *Theory of Technical Change and Economic Invariance: Application of Lie Groups.* [25.2]

Sauer, C.H. *Computer Systems Performance Modeling.* [19.3]

Saunders, Hal M. *Mathematics for the Trades: A Guided Approach.* [16.2]

** Saupe, Dietmar. *The Science of Fractal Images.* [8.5]

Savage, I. Richard. *Index to Volumes 1–31, 1930–1960: The Annals of Mathematical Statistics.* [2.5]

Savage, John E. *The Mystical Machine: Issues and Ideas in Computing.* [22.1]

Savage, Katherine. *Nursing Simplified: Math Logic.* [16.3]

Savage, Leonard J. *Inequalities for Stochastic Processes (How to Gamble If You Must).* [20.4]

* Savitch, Walter J. *Abstract Machines and Grammars.* [22.15]

Sawyer, W.W. *Introducing Mathematics.* [1.3]

*** ———. *What is Calculus About?* [6.5]

* Scarf, Herbert. *The Computation of Economic Equilibria.* [25.2]

** Schaaf, William L. *A Bibliography of Recreational Mathematics.* [2.4]

———. *Mathematics and Science: An Adventure in Postage Stamps.* [2.4]

———. *The High School Mathematics Library.* [2.4]

Schaefer, Barbara K. *Using the Mathematical Literature: A Practical Guide.* [2.4]

Schaffer, William A. *Cost-Benefit Analysis: A Handbook.* [25.2]

* Scharlau, Winfried. *From Fermat to Minkowski: Lectures on the Theory of Numbers and Its Historical Development.* [11.2]

Schattschneider, Doris J. *M.C. Escher Kaleidocycles.* [4.6]

** ———. *Visions of Symmetry: Notebooks, Periodic Drawings, and Related Work of M.C. Escher.* [14.5]

Scheaffer, Richard L. *Probability and Statistics for Engineers.* [21.4]

** ———. *The Art and Techniques of Simulation.* [20.1]

** ———. *Elementary Survey Sampling.* [21.13]

** ———. *Exploring Probability.* [20.1]

** ———. *Introduction to Probability and Its Applications.* [20.2]

———. *Mathematical Statistics with Applications.* [21.4]

* Schell, Frank R. *Practical Problems in Mathematics for Welders.* [16.2]

Schendel, U. *Introduction to Numerical Methods for Parallel Computers.* [18.6]

** Schey, H.M. *Div, Grad, Curl, and All That: An Informal Text on Vector Calculus.* [6.4]

* Schiffer, M.M. *The Role of Mathematics in Science.* [1.3]

Schikhof, W.H. *Ultrametric Calculus: An Introduction to p-adic Analysis.* [11.8]

* Schlesinger, B. *Thinking Through Mathematics: Fostering Inquiry and Communication in Mathematics Classrooms.* [5.6]

* Schmidt, Eckart. *n-gons.* [13.13]

Schmidt, J. William. *Foundations of Analysis in Operations Research.* [19.1]

* ———. *Mathematical Foundations for Management Science and Systems Analysis.* [19.1]

** Schneider, David I. *Finite Mathematics and Its Applications.* [10.2]

* ———. *Calculus and Its Applications.* [6.3]

Schnibben, George E. *Infinite Algebraic Extensions of Finite Fields.* [13.7]

* Schoen, Harold L. *Estimation and Mental Computation: 1986 Yearbook.* [5.6]

Schoenberg, I.J. *Mathematical Time Exposures.* [1.2]

* Schoenfeld, Alan H. *A Sourcebook for College Mathematics Teaching.* [5.9]

* ———. *Cognitive Science and Mathematics Education.* [5.3]

*** ———. *Mathematical Problem Solving.* [5.11]

* ———. *Problem Solving in the Mathematics Curriculum: A Report, Recommendations, and an Annotated Bibliography.* [5.11]

** Schrage, Linus E. *A Guide to Simulation.* [19.9]

* Schrijver, Alexander. *Theory of Linear and Integer Programming.* [19.7]

** Schroeder, Manfred R. *Fractals, Chaos, Power Laws: Minutes from an Infinite Paradise.* [8.5]

** ———. *Number Theory in Science and Communication with Applications in Cryptography, Physics,*

* Shubik, Martin. *Game Theory in the Social Sciences.* [19.4]

Shubnikov, A.V. *Symmetry in Science and Art.* [14.5]

*** Shulte, Albert P. *New Directions for Elementary School Mathematics: 1989 Yearbook.* [5.7]

* ———. *Teaching Statistics and Probability: 1981 Yearbook.* [5.6]

** ———. *Learning and Teaching Geometry, K–12: 1987 Yearbook.* [5.6]

*** ———. *The Ideas of Algebra, K–12: 1988 Yearbook.* [5.8]

Shumate, K. *Understanding Ada with Abstract Data Types.* [22.7]

Shumway, Richard J. *Research in Mathematics Education.* [5.12]

* Shute, Charles. *The Rhind Mathematical Papyrus: An Ancient Egyptian Text.* [3.4]

** Siegel, Andrew F. *Counterexamples in Probability and Statistics.* [20.8]

** ———. *Statistics and Data Analysis: An Introduction.* [21.2]

** Siegel, Martha J. *Finite Mathematics and Its Applications.* [10.2]

Siegmund, David. *Great Expectations: The Theory of Optimal Stopping.* [20.4]

* Sierpiński, W. *250 Problems in Elementary Number Theory.* [11.3]

———. *A Selection of Problems in the Theory of Numbers.* [11.9]

* ———. *Elementary Theory of Numbers.* [11.3]

———. *General Topology.* [15.1]

———. *Theory of Numbers.* [11.3]

* Siewiorek, Daniel P. *Computer Structures: Principles and Examples.* [22.17]

Sigler, L.E. *Leonardo Pisano Fibonacci: The Book of Squares.* [3.4]

Sikorski, R. *Boolean Algebras.* [13.13]

* Silberschatz, Abraham. *Operating System Concepts.* [22.10]

———. *Database System Concepts.* [22.5]

** Silver, Edward A. *The Teaching and Assessing of Mathematical Problem Solving.* [5.12]

———. *Teaching and Learning Mathematical Problem Solving: Multiple Research Perspectives.* [5.11]

* ———. *Thinking Through Mathematics: Fostering Inquiry and Communication in Mathematics Classrooms.* [5.6]

* Silverman, David L. *Your Move.* [4.3]

Silverman, Joseph H. *The Arithmetic of Elliptic Curves.* [11.7]

Silvester, John R. *Introduction to Algebraic K-Theory.* [13.13]

Silvey, Linda. *Mathematics for the Middle Grades (5–9): 1982 Yearbook.* [5.7]

* Simmonds, James G. *A Brief on Tensor Analysis.* [6.4]

** Simmons, Donald M. *Nonlinear Programming for Operations Research.* [19.6]

** Simmons, George F. *Differential Equations with Applications and Historical Notes.* [7.1]

* ———. *Calculus with Analytic Geometry.* [6.3]

* ———. *Introduction to Topology and Modern Analysis.* [15.1]

———. *Precalculus Mathematics in a Nutshell: Geometry, Algebra, Trigonometry.* [6.2]

* Simon, Barry. *Methods of Modern Mathematical Physics.* [24.3]

Simon, William. *Mathematical Magic.* [4.6]

Simpson, F. Morgan. *Activities for Junior High School and Middle School Mathematics.* [5.8]

Sims, Charles C. *Abstract Algebra: A Computational Approach.* [13.2]

Sincovec, R. *Programming in ADA.* [22.7]

*** Singer, B.B. *Basic Mathematics for Electricity and Electronics.* [16.5]

*** Singer, I.M. *Lecture Notes on Elementary Topology and Geometry.* [15.4]

Singmaster, David. *Handbook of Cubik Math.* [4.2]

Sinha, Naresh K. *Microcomputer-Based Numerical Methods for Science and Engineering.* [18.6]

* Siret, Y. *Computer Algebra: Systems and Algorithms for Algebraic Computation.* [22.8]

Sirovich, L. *Introduction to Applied Mathematics.* [24.2]

Skelley, Esther G. *Medications and Mathematics for the Nurse.* [16.3]

** Skemp, Richard R. *The Psychology of Learning Mathematics.* [5.3]

Skorokhod, A.V. *Studies in the Theory of Random Processes.* [20.4]

Skowronski, Janislaw M. *Renewable Resource Management.* [23.2]

** Slater, Jeffrey. *Practical Business Math Procedures.* [17.1]

** Sloane, N.J.A. *A Handbook of Integer Sequences.* [2.2]

———. *A Short Course on Error Correcting Codes.* [10.6]

** ———. *The Theory of Error-Correcting Codes.* [10.6]

* Slovic, Paul. *Judgment Under Uncertainty: Heuristics and Biases.* [20.1]

*** Smale, Stephen. *Differential Equations, Dynamical Systems, and Linear Algebra.* [7.2]

Small, Charles. *Introduction to Homological Methods in Commutative Rings.* [13.9]

Small, Donald B. *Calculus: An Integrated Approach.* [6.3]

———. *Explorations in Calculus with a Computer Algebra System.* [6.3]

Small, Lance W. *Reviews in Ring Theory, 1960–1984.* [2.6]

———. *Noetherian Rings and Their Applications.* [13.6]

Smart, James R. *Mathematics for the Middle Grades (5–9): 1982 Yearbook.* [5.7]

———. *Modern Geometries.* [14.2]

Smith, D. *A Transition to Advanced Mathematics.* [8.1]

Smith, David A. *Computers and Mathematics: The Use of Computers in Undergraduate Instruction.* [5.9]

Swartzlander, Earl E., Jr. *Computer Arithmetic.* [22.8]
Swerdlow, N.M. *Mathematical Astronomy in Copernicus's De Revolutionibus.* [3.4]
* Swetz, Frank J. *Was Pythagoras Chinese? An Examination of Right Triangle Theory in Ancient China.* [3.5]
* ———. *Capitalism and Arithmetic: The New Math of the 15th Century.* [3.4]
** Swift, Jim. *Exploring Surveys and Information From Samples.* [21.1]
** ———. *The Art and Techniques of Simulation.* [20.1]
* Swokowski, Earl W. *Algebra and Trigonometry with Analytic Geometry.* [6.2]
———. *Calculus.* [6.3]
Syski, R. *Introduction to Congestion Theory in Telephone Systems.* [19.3]
———. *Random Processes: A First Look.* [20.4]
** Szebehely, Victor G. *Adventures in Celestial Mechanics: A First Course in the Theory of Orbits.* [24.8]
* Szegő, Gábor. *Orthogonal Polynomials.* [8.4]
*** ———. *Problems and Theorems in Analysis.* [8.3]
———. *Toeplitz Forms and Their Applications.* [8.3]
** Székely, Gábor J. *Paradoxes in Probability Theory and Mathematical Statistics.* [20.8]
Szeminska, A. *The Child's Conception of Geometry.* [5.3]
* Szmielew, Wanda. *From Affine to Euclidean Geometry: An Axiomatic Approach.* [14.4]

T

Taha, Hamdy A. *Integer Programming: Theory, Applications, and Computations.* [19.7]
* Takeuti, Gaisi. *Proof Theory.* [9.9]
Tall, David. *Algebraic Number Theory.* [11.5]
* Tan, S.T. *Applied Finite Mathematics.* [10.2]
* Tanenbaum, Andrew S. *Operating Systems: Design and Implementation.* [22.10]
*** ———. *Structured Computer Organization.* [22.17]
** Tanis, Elliot A. *Probability and Statistical Inference.* [21.4]
*** Tanur, Judith M. *Statistics: A Guide to the Unknown.* [21.1]
———. *International Encyclopedia of Statistics.* [2.3]
*** ———. *Statistics: A Guide to Business and Economics.* [17.3]
*** ———. *Statistics: A Guide to Political and Social Issues.* [25.1]
** ———. *Statistics: A Guide to the Study of the Biological and Health Sciences.* [23.1]
* Tarapov, I.E. *Vector and Tensor Analysis with Applications.* [24.11]
Tarjan, Robert E. *Notes on Introductory Combinatorics.* [10.4]
Tarski, Alfred. *Cylindric Algebras.* [9.9]
* ———. *Logic, Semantics, Metamathematics.* [9.3]
* ———. *Ordinal Algebras.* [9.9]
Tarwater, J. Dalton. *American Mathematical Heritage: Algebra and Applied Mathematics.* [3.6]
———. *Men and Institutions in American Mathematics.* [3.6]
Tate, John. *Class Field Theory.* [11.5]
Taub, A.H. *Studies in Applied Mathematics.* [24.1]
Tautu, P. *Biological Growth and Spread: Mathematical Theories and Applications.* [23.8]
———. *Stochastic Processes and Applications in Biology and Medicine.* [23.7]
* Taylor, Angus E. *Introduction to Functional Analysis.* [8.8]
———. *Advanced Calculus.* [6.4]
———. *General Theory of Functions and Integration.* [8.3]
* Taylor, Howard M. *An Introduction to Stochastic Modeling.* [20.4]
* ———. *A First Course in Stochastic Processes.* [20.4]
———. *A Second Course in Stochastic Processes.* [20.4]
* Taylor, Ross. *Professional Development for Teachers of Mathematics: A Handbook.* [5.6]
* Taylor, S.J. *Introduction to Measure with Probability.* [20.6]
Teeters, J.L. *Creating Escher-Type Drawings.* [14.5]
* Teicher, Henry. *Probability Theory: Independence, Interchangeability, Martingales.* [20.3]
Temple, George. *100 Years of Mathematics: A Personal Viewpoint.* [3.1]
* Terras, Audrey. *Harmonic Analysis on Symmetric Spaces and Applications.* [8.13]
Terry, Patrick D. *Programming Language Translation: A Practical Approach.* [22.16]
* Thie, Paul R. *An Introduction to Linear Programming and Game Theory.* [19.5]
* Thisted, Ronald A. *Elements of Statistical Computing: Numerical Computation.* [21.8]
Thoe, Dale W. *Introduction to Partial Differential Equations with Applications.* [7.4]
* Thom, René. *Mathematical Models of Morphogenesis.* [23.6]
———. *Structural Stability and Morphogenesis: An Outline of a General Theory of Models.* [23.6]
*** Thomas, George B., Jr. *Calculus and Analytic Geometry.* [6.3]
* ———. *Calculus.* [6.3]
Thomas, L.C. *Games: Theory and Applications.* [19.4]
Thomas, Melvin. *Ideas from the Arithmetic Teacher, Grades 1–4: Primary.* [5.7]
* ———. *Ideas from the Arithmetic Teacher, Grades 6–8: Middle School.* [5.8]
* Thomas, W. *Mathematical Logic.* [9.3]
* Thompson, Colin J. *Mathematical Statistical Mechanics.* [24.13]
*** Thompson, Gerald L. *Introduction to Finite Mathematics.* [10.2]
* Thompson, James R. *Empirical Model Building.* [19.2]
* Thompson, J.M.T. *Instabilities and Catastrophes in Science and Engineering.* [24.3]
** Thompson, Maynard. *Life Science Models.* [23.1]
* ———. *Finite Mathematics.* [10.2]

Tsay, Jia-Yeong. *Statistics in the Pharmaceutical Industry.* [21.16]
Tu, Pierre N.V. *Introductory Optimization Dynamics: Optimal Control with Economics and Management Science Applications.* [19.10]
* Tucker, Alan. *A Unified Introduction to Linear Algebra: Models, Methods, and Theory.* [12.1]
** ———. *Applied Combinatorics.* [10.3]
* Tucker, Allen B. *Fundamentals of Computing I: Logic, Problem-Solving, Programs, and Computers.* [22.3]
———. *Programming Languages.* [22.6]
* Tucker, Howard G. *A Graduate Course in Probability.* [20.3]
** Tucker, Thomas W. *Priming the Calculus Pump: Innovations and Resources.* [5.9]
———. *Topological Graph Theory.* [10.5]
* Tuckwell, Henry C. *Elementary Applications of Probability Theory.* [20.2]
** Tufte, Edward R. *Envisioning Information.* [21.1]
*** ———. *The Visual Display of Quantitative Information.* [21.1]
* Tukey, John W. *Exploring Data Tables, Trends, and Shapes.* [21.6]
*** ———. *Understanding Robust and Exploratory Data Analysis.* [21.6]
** ———. *Data Analysis and Regression: A Second Course in Statistics.* [21.9]
———. *Exploratory Data Analysis.* [21.6]
———. *Index to Statistics and Probability: Permuted Titles.* [2.5]
Tuller, Anita. *A Modern Introduction to Geometries.* [14.2]
Tutte, W.T. *Graph Theory.* [10.5]
Tversky, Amos. *Mathematical Psychology: An Elementary Introduction.* [25.4]
* ———. *Judgment Under Uncertainty: Heuristics and Biases.* [20.1]
Twizell, E.H. *Computational Methods for Partial Differential Equations.* [18.3]

U

* Uhl, J.J., Jr. *The Mathematics of Nonlinear Programming.* [19.6]
** Ulam, S.M. *Adventures of a Mathematician.* [3.2]
*** Ullman, Jeffrey D. *Compilers: Principles, Techniques, and Tools.* [22.16]
*** ———. *Data Structures and Algorithms.* [22.4]
*** ———. *Introduction to Automata Theory, Languages, and Computation.* [22.15]
* ———. *Principals of Database and Knowledge-based Systems.* [22.5]
*** ———. *The Design and Analysis of Computer Algorithms.* [22.8]
Urbach, Peter. *Scientific Reasoning: The Bayesian Approach.* [21.7]
* Usiskin, Zalman. *Advanced Algebra with Transformations and Applications.* [6.2]
* Uspenskii, V.A. *Post's Machine.* [9.6]
———. *Problems in the Theory of Numbers.* [11.9]
** Uspensky, J.V. *Elementary Number Theory.* [11.1]

V

Vaisman, Izu. *Foundations of Three-Dimensional Euclidean Geometry.* [14.4]
Vajda, Steven. *Algebra.* [13.4]
———. *Analysis.* [24.1]
* ———. *Geometry and Combinatorics.* [14.9]
* van Daal, J. *On the Height of a Railway Bridge.* [19.9]
van Dalen, Dirk. *Sets and Integration: An Outline of the Development.* [3.9]
———. *Brouwer's Cambridge Lectures on Intuitionism.* [9.8]
** ———. *Constructivism in Mathematics: An Introduction.* [9.8]
* Van de Walle, John A. *Elementary School Mathematics: Teaching Developmentally.* [5.7]
van der Ploeg, Frederick. *Mathematical Methods in Economics.* [25.2]
** van der Waerden, B.L. *A History of Algebra: From al-Khwārizmī to Emmy Noether.* [3.8]
*** ———. *Algebra.* [13.4]
———. *Geometry and Algebra in Ancient Civilizations.* [3.5]
** ———. *Science Awakening.* [3.5]
van der Werff, T.J. *Mathematical Models of the Dynamics of the Human Eye.* [23.6]
* van Doeland, F. *On the Height of a Railway Bridge.* [19.9]
* Van Heijenoort, Jean. *From Frege to Gödel: A Source Book in Mathematical Logic, 1879–1931.* [3.3]
van Lint, J.H. *Graphs, Codes and Designs.* [10.6]
** Van Loan, Charles F. *Handbook for Matrix Computations.* [18.4]
*** ———. *Matrix Computations.* [18.4]
van Stigt, Walter P. *Brouwer's Intuitionism.* [3.2]
Vanden Eynden, Charles. *Applied Mathematics for the Management, Life, and Social Sciences.* [17.2]
———. *Finite Mathematics.* [10.2]
Vandergraft, James S. *Introduction to Numerical Computations.* [18.2]
* Varberg, Dale E. *Faces of Mathematics: An Introductory Course for College Students.* [1.4]
Varga, Richard S. *Matrix Iterative Analysis.* [18.4]
Veblen, Oswald. *The Foundations of Differential Geometry.* [14.10]
* ———. *Projective Geometry.* [14.7]
Velleman, Paul F. *Applications, Basics, and Computing of Exploratory Data Analysis.* [21.6]
Vemuri, V. *Digital Computer Treatment of Partial Differential Equations.* [18.3]
Venn, John. *Symbolic Logic.* [9.2]
———. *The Principles of Inductive Logic.* [9.2]
Vergnaud, G. *L'enfant, la mathématique et la réalité.* [5.3]
Verhulst, F. *Nonlinear Differential Equations and Dynamical Systems.* [7.3]

** Watt, Alan. *Fundamentals of Three Dimensional Computer Graphics.* [22.18]
Watt, J.M. *Modern Numerical Methods for Ordinary Differential Equations.* [18.3]
* Watts, Donald G. *Nonlinear Regression Analysis and Its Applications.* [21.9]
Weaver, Mabel E. *Programmed Mathematics of Drugs and Solutions.* [16.3]
*** Weaver, Warren. *Lady Luck: The Theory of Probability.* [20.1]
*** ———. *The Mathematical Theory of Communication.* [24.12]
Webber, G. *Basic Concepts of Mathematics.* [6.1]
Webre, Neil W. *Data Structures.* [22.4]
Wechsler, Judith. *On Aesthetics in Science.* [1.6]
Wedderburn, J.H.M. *Lectures on Matrices.* [12.3]
** Weeks, Jeffrey R. *The Shape of Space: How to Visualize Surfaces and Three-Dimensional Manifolds.* [15.2]
Weil, André. *Basic Number Theory.* [11.5]
———. *Elliptic Functions According to Eisenstein and Kronecker.* [11.7]
*** ———. *Number Theory: An Approach Through History From Hammurapi to Legendre.* [11.2]
*** ———. *Number Theory for Beginners.* [11.1]
** Weinberger, Hans F. *A First Course in Partial Differential Equations with Complex Variables and Transform Methods.* [7.4]
* ———. *Maximum Principles in Differential Equations.* [7.4]
Weinberger, Peter J. *The AWK Programming Language.* [22.10]
* Weinstein, Alan. *Calculus III.* [6.4]
* ———. *Calculus.* [6.3]
Weinstein, Michael. *Examples of Groups.* [13.5]
Weinstock, Robert. *Calculus of Variations with Applications to Physics and Engineering.* [8.10]
* Weir, Maurice D. *A First Course in Mathematical Modeling.* [19.2]
*** Weisberg, Sanford. *Applied Linear Regression.* [21.9]
* ———. *Residuals and Influence in Regression.* [21.9]
** Weiss, Eric A. *A Computer Science Reader: Selections from Abacus.* [22.1]
** Weiss, G. *Introduction to Fourier Analysis on Euclidean Spaces.* [8.4]
Weiss, Max L. *Geometry and Convexity: A Study in Mathematical Methods.* [14.9]
Weissglass, Julian. *Exploring Elementary Mathematics: A Small-Group Approach for Teaching.* [5.7]
** Weizenbaum, Joseph. *Computer Power and Human Reason: From Judgment to Calculation.* [22.2]
* Weldon, E.J., Jr. *Error-Correcting Codes.* [10.6]
* Wells, Charles. *Category Theory for Computing Science.* [13.10]
Wells, David. *The Penguin Dictionary of Curious and Interesting Geometry.* [2.2]
* Wells, R.O., Jr. *Mathematics in Civilization.* [1.4]
Wells, Timothy D. *A Structured Approach to Building Programs: Pascal.* [22.14]
Welsch, Roy E. *Regression Diagnostics: Identifying Influential Data and Sources of Collinearity.* [21.9]
* Welsh, D.J.A. *Matroid Theory.* [10.7]
* Welsh, Dominic. *Codes and Cryptography.* [10.6]
Welsh, Jim. *Sequential Program Structures.* [22.4]
* Wenninger, Magnus J. *Polyhedron Models.* [14.5]
———. *Spherical Models.* [14.5]
Wermer, John. *Banach Algebras and Several Complex Variables.* [8.7]
** ———. *Linear Algebra Through Geometry.* [12.1]
Wesner, Terry H. *Essentials of Technical Mathematics.* [16.1]
** West, Beverly H. *Differential Equations: A Dynamical Systems Approach.* [7.1]
Westbury, Ian. *The IEA Study of Mathematics I: Analysis of Mathematics Curricula.* [5.13]
** Westfall, Richard S. *Never at Rest: A Biography of Isaac Newton.* [3.2]
Wetherill, G. Barrie. *Intermediate Statistical Methods.* [21.4]
Wetterling, W. *Optimization Problems.* [19.5]
*** Wexelblat, Richard L. *History of Programming Languages.* [22.6]
Weyl, F. Joachim. *The Spirit and the Uses of the Mathematical Sciences.* [1.1]
*** Weyl, Hermann. *Symmetry.* [14.5]
* ———. *The Classical Groups: Their Invariants and Representatives.* [13.5]
* ———. *The Concept of a Riemann Surface.* [8.7]
Weyuker, Elaine J. *Computability, Complexity, and Languages: Fundamentals of Theoretical Computer Science.* [22.9]
* Wheatley, Patrick O. *Applied Numerical Analysis.* [18.1]
Wheeden, Richard L. *Measure and Integral: An Introduction to Real Analysis.* [8.2]
*** Wheeler, John Archibald. *Gravitation.* [24.7]
Wheeler, Margariete M. *Mathematics Library: Elementary and Junior High School.* [2.4]
White, Arthur T. *Graphs, Groups and Surfaces.* [10.5]
** White, Dennis. *Constructive Combinatorics.* [10.3]
White, John T. *Men and Institutions in American Mathematics.* [3.6]
White, Neil. *Theory of Matroids.* [10.7]
Whitehead, Alfred North. *An Introduction to Mathematics.* [1.2]
Whitehead, J.H.C. *The Foundations of Differential Geometry.* [14.10]
Whitemore, Hugh. *Breaking the Code.* [3.2]
Whiteside, D.T. *The Mathematical Papers of Isaac Newton.* [3.4]
* Whitham, G. *Linear and Nonlinear Waves.* [7.4]
*** Whittaker, Edmund T. *A Course of Modern Analysis.* [8.7]
Whittaker, R.H. *Niche: Theory and Application.* [23.2]
———. *Communities and Ecosystems.* [23.2]
Whittle, Peter. *Optimization Over Time: Programming and Stochastic Control.* [19.3]
** Wichern, Dean W. *Applied Multivariate Statistical Analysis.* [21.10]
———. *Intermediate Business Statistics.* [17.3]
* Widder, David V. *Advanced Calculus.* [6.4]
———. *An Introduction to Transform Theory.* [8.4]

Wright, Margaret H. *Practical Optimization.* [19.6]
Wulf, William A. *Fundamental Structures of Computer Science.* [22.3]
* Wussing, Hans. *The Genesis of the Abstract Group Concept.* [3.8]
Wylie, Clarence R. *Advanced Engineering Mathematics.* [24.2]
———. *Introduction to Projective Geometry.* [14.7]

Y

Yager, R.R. *Fuzzy Sets and Possibility Theory.* [20.7]
Yaglom, A.M. *Challenging Mathematical Programs with Elementary Solutions.* [4.4]
* ———. *Probability and Information.* [20.8]
* Yaglom, I.M. *Convex Figures.* [14.9]
———. *A Simple Non-Euclidean Geometry and Its Physical Basis.* [14.4]
———. *Challenging Mathematical Programs with Elementary Solutions.* [4.4]
———. *Felix Klein and Sophus Lie: Evolution of the Idea of Symmetry in the Nineteenth Century.* [3.10]
** ———. *Geometric Transformations.* [14.6]
* ———. *Probability and Information.* [20.8]
Yates, Robert C. *Curves and Their Properties.* [14.12]
* ———. *The Trisection Problem.* [14.4]
Yoder, Joella G. *Unrolling Time: Christiaan Huygens and the Mathematization of Nature.* [3.13]
** Yosida, Kôsaku. *Functional Analysis.* [8.8]
* ———. *Lectures on Differential and Integral Equations.* [7.6]
Youden, W.J. *Experimentation and Measurement.* [21.3]
* ———. *Risk, Choice, and Prediction: An Introduction to Experimentation.* [21.3]
Young, David M. *A Survey of Numerical Mathematics.* [18.2]
———. *Iterative Solution of Large Linear Systems.* [18.4]
Young, Gail S. *The Future of College Mathematics.* [5.9]
Young, Grace Chisholm. *The Theory of Sets of Points.* [3.4]
*** Young, H. Peyton. *Fair Representation: Meeting the Ideal of One Man, One Vote.* [25.3]
* Young, John Wesley. *Projective Geometry.* [14.7]
* ———. *Projective Geometry.* [14.7]
Young, L.C. *Lectures on the Calculus of Variations and Optimal Control Theory.* [19.10]
Young, Nicholas. *An Introduction to Hilbert Space.* [8.8]
Young, William H. *The Theory of Sets of Points.* [3.4]

Z

Zachmanoglou, E.C. *Introduction to Partial Differential Equations with Applications.* [7.4]
Zadeh, Lotfi. *Linear Systems Theory.* [24.3]
*** Zariski, Oscar. *Commutative Algebra.* [13.8]
Zarrella, John. *Language Translators.* [22.16]
* Zaslavsky, Claudia. *Africa Counts: Number and Pattern in African Culture.* [5.4]
Zassenhaus, H. *Algorithmic Algebraic Number Theory.* [11.5]
* Zauderer, Erich. *Partial Differential Equations of Applied Mathematics.* [7.4]
Zeeman, E.C. *Catastrophe Theory: Selected Papers, 1972–1977.* [19.2]
* Zehnwirth, B. *Introductory Statistics with Applications in General Insurance.* [17.3]
Zerna, W. *Theoretical Elasticity.* [24.4]
Ziegler, J. *The Golden Age of Theoretical Ecology.* [23.2]
*** Ziegler, Michael R. *Applied Mathematics for Business, Economics, Life Sciences, and Social Sciences.* [17.2]
* ———. *Finite Mathematics for Management, Life, and Social Sciences.* [17.2]
Zienkiewicz, O.C. *The Finite Element Method.* [18.7]
Zieschang, Heiner. *Knots.* [15.2]
** Zill, Dennis G. *A First Course in Differential Equations with Applications.* [7.1]
Zima, P. *Mathematics of Finance.* [17.5]
* Zimmermann, Walter. *Visualization in Teaching and Learning Mathematics.* [5.9]
Zinner, Ernst. *Regiomontanus: His Life and Work.* [3.5]
Zippin, Leo. *Topological Transformation Groups.* [8.13]
Zitarelli, David E. *The Ascent of Mathematics.* [1.4]
*** Zuckerman, Herbert S. *An Introduction to the Theory of Numbers.* [11.1]
Zweng, Marilyn J. *Proceedings of the Fourth International Congress on Mathematical Education.* [5.6]
———. *Computers in Mathematics Education: 1984 Yearbook.* [5.14]
* ———. *Estimation and Mental Computation: 1986 Yearbook.* [5.6]
** ———. *The Secondary School Mathematics Curriculum: 1985 Yearbook.* [5.8]
** Zwillinger, Daniel. *Handbook of Differential Equations.* [2.2]
** Zygmund, Antoni. *Analytic Functions.* [8.7]
———. *Measure and Integral: An Introduction to Real Analysis.* [8.2]
———. *Trigonometric Series.* [8.4]